Logic and Knowledge Representation

an introduction for systems analysts

Logic and Knowledge Representation

an introduction for systems analysts

Alwyn Jones
School of Informatics
The City University

Pitman

PITMAN PUBLISHING
128 Long Acre, London WC2E 9AN

A Division of Longman Group UK Limited

© A Jones 1991

First published in Great Britain 1991

British Library Cataloguing in Publication Data

Jones, Alwyn
   Logic and knowledge representation: an introduction for
systems analysts.
   1. Systems analysis. Applications of logic
   I. Title
   003.01

   ISBN 0-273-03150-3

ISBN 0 273 03150 3

Printed and bound in Great Britain

Contents

x

A ➤ B with love

1 The Logical Systems Analyst

1.1 Objectives

1.1.1 Whom are we addressing?

The epithet Systems Analyst seems to have become settled, in the UK at least, as the job-name of a person whose professional task it is to develop working computer-based information systems for business and industry. We shall not here attempt to produce refined definitions of that job, but appeal to common understanding and to the common sense of the reader in interpreting whom it is our intention to address in this book.

However, one or two points do need to be made. The first concerns the question of how specialised the job is understood to be if there are additional qualifying adjectives added to the title, such as, for example, maintenance systems analyst or design analyst or analyst programmer. The intention is to embrace all these 'specialist' systems analysts as our audience. In the same way there is no intention to exclude systems analysts who may be employed within particular industries or business areas, or those who find themselves specialising in the application of particular types of software and so on.

This much should be fairly obvious. However, rather less obvious is the question as to whether the uncomfortably termed 'end-user' should be accepted as at least a potential or embryonic systems analyst. (The end-user is in fact any person who comes into contact with the day-to-day use of computing facilities. For some this means considerable naivety about the inner workings of the machine and/or the software; for others it means a fairly deep and increasing knowledge of computing and its modes of application.) Much will depend here upon the actual experience and understanding of the individual concerned. What we say in due course concerning the value to the systems analyst of understanding the basics of logic and of knowledge representation and processing could very well apply to the enquiring mind of an end-user, but without perhaps being able to justify that such understanding is actually vitally needed. As we shall have to admit, even many systems analysts may get away with having relatively little understanding in this area, but such persons will soon not be able to claim to be fully competent systems analysts in the contemporary scene.

Another point we need to address in view of the book's title and its patent scope is the rather thornier issue as to whether the systems analyst is to be expected to include knowledge engineering as one of the many skills listed as being a required ability. Whatever the views of the theorists and purists about the need to separate out knowledge engineering as being far too specialised a job for the average systems analyst, the commercial reality is that at least some people within traditional data processing departments will be expected to be able to perform such work and most systems analysts expected to have intelligent advice to offer. It is true that we do at present have a trickle of graduates who will be comfortable with the title Knowledge Engineer but it is only a trickle. And these people are more than likely to have their focus of attention on the specialism itself; what then happens to the growing need for systems analysis to use knowledge-based tools in their own analysis and design work?

Finally, because the book is on the whole written with the assumption in mind that the systems analyst is often not an educated mathematician, there is the possibility that it may well be of use to the non-systems analyst who is seeking for whatever reason to understand, or at least come to terms with, the basics of logic and its relevance in this age of growing concern with so-called artificial intelligence. Such readers are welcome, though they may find that the examples and exercises have a particular systems analysis flavour for them which they may not always be able to fully empathise with.

1.1.2 Who says the systems analyst needs to be logical?

We would certainly not want to say that the systems analyst needs to be purely logical to be successful. Indeed a crucial aspect of the work is the need at various times to be creative, imaginative, socially adept, tactful and determined. None of these qualities are necessarily enhanced by having a purely logical approach to the work; indeed logic may even at times hinder the ability to excel at any one of them.

The exemplary systems analyst belongs to that group of designers who desperately want what they have designed to work. Not just to work but to work effectively, efficiently and to the delight and admiration of the users. When one looks at the horrendous complexity of some of these systems one might also add laconically that the systems analyst also belongs to that group of designers who live in a dream-world! It cannot be denied that the ambition most certainly incorporates the wish to get working what is seriously a technically based system.

Now, while it is technically based, the system also has the unavoidable requirement that it works properly in the context of essentially human activity. And not simply any old activity but that which is only meaningful when engaging fairly intimately with the human mind—what we may call cognitive or cognate activity. It would not otherwise be an information system, that is to say a system which requires at some point to involve the attention of human intellect. If we wish to find some way of distinguishing data from information, we could say that data becomes information the moment it is perceived and mentally received by a human being.

So the systems analyst is not designing systems which are required only to be passively admired by others, but systems whose esteem has to survive the intimate involvement of its evaluator. Hence the recognition at last in the late eighties that the human–computer interface (HCI) has to be designed with great skill and sympathy for human behaviour, especially the cognate kind of behaviour. And yet at the heart of all this the system is still a technical one. It is instructive at some early stage of a major project to pause and contemplate that at the end of the story there will have to be a dynamic system of moving electronic binary digits which must somehow successfully model a significant proportion of the complex processes of the current human decision-making processes, and probably other types of process besides. This bridge from bits to human activity system is an awesome idea to hold on to and yet all around us in business, in industry and, increasingly, in recreation there are such systems in place and working well.

Apart from the wealth of creativity, imagination, social intelligence, tactfulness and determination that must have gone into this complex bridge-building, can any one seriously deny that there must also have been a considerable amount of logical deduction and reasoning on the part of the systems analyst, to say nothing of that on the part of the programmers and other technicians allied to the project? From the hard-edged reasoning about such things as data formats and file structures to the more intuitive kind related to modelling the probable thought behaviour of the users, the demand for logical reasoning on the part of the designer is surely irrefutable. And as the data processing content becomes ever more enhanced with information and knowledge content, so the argument must gain ever more strength.

1.1.3 The systems analyst profession today

The computer scientist, newly qualified and embarking on a systems analysis career, may look through this book and declare that it doesn't look very mathematical for a text on logic and knowledge processing. A systems analyst of some years experience may look at it and express fears that it looks too mathematical for a person who came into the profession via some definitely non-mathematical pathways. An old problem.

However, let us make clear at the outset that the book is aimed more at the latter then the former, while hoping at the same time to instruct and entertain the bulk of those in between the above extremes (if 'extremes' is what they are). The salient point in the author's view is that a very significant portion of any profession is at any one time educated up to a standard set some way into the past. Technical training courses do of course help to keep practitioners up to date in skills and in selective narrow seams of knowledge, but their deeper understanding of the current state of knowledge relating to the subject of their profession becomes ever more fragile as important underlying principles seem increasingly remote and therefore difficult to grasp.

This situation definitely seems to have happened in the case of the growth of knowledge processing concepts and techniques which are of relevance to the work of the systems analyst. Thus, if nothing else, this book is most certainly dedicated to enlightening today's practising systems analysts.

However, another relevant point here is that a great number of neophyte systems analysts are being recruited from courses of education which have not touched upon the mathematics or even the ideas of knowledge processing, and they too should find help here. There are quite a number of conversion post-graduate courses which carry students who find they need help in this area from other than the mathematicians and highly technical experts.

The reference above to 'important underlying principles' may threaten to require mathematical understanding from the reader in spite of what has been said. Certainly some acceptance of the symbolic approach to the description of reasoning is essential but we shall attempt to use such methods only in so far as they are genuinely needed to help in the understanding of principles. Also illustration and interpretation will be related to what is hoped will be familiar types of problem in the practical context. Although we do not subscribe to the school of thought that suggests that systems analysts must first cut their teeth as skilled programmers, it is felt not unreasonable that the reader should be familiar with the basic principles of programming.

1.1.4 How the demands are changing

Let us look a little more closely at the current developments in systems analysis which are relevant to our theme. Our theme, remember, is concerned with logic and knowledge representation and processing (which we have so far deliberately avoided defining, and shall continue to do for the time being).

One radical viewpoint about systems analysis is that it will one day soon vanish! This, it is argued, is because from the users' arena there is emerging the intelligent and fearless personal computer software user who can turn a hand to anything so long as there is sophisticated software to show the way—this is the shining prospect of true end-user computing. From the technical fraternity come the fourth generation language users, using the power of their tools to sweep complete applications into existence with a few lines of parameterised code. Finally, there are the application package vendors who will fit a new system into the organisation as neatly and cleanly as a domestic washing-machine installer. Between them they leave no room for the designer of customised and expensive systems which is how this argument generally views the systems analyst.

Not only are each of these types idealised but the picture fails to take into account the reality of the number of systems analysts actually in employment and being expected to deliver in a wide spectrum of ways which the 'designer of customised systems' view fails to recognise. (This last point is rather like the situation with respect to expected programming skills as demanded by business. The language COBOL was pronounced by purists to be a dying language as early as the late 70s due to the growth of fourth generation languages; in February 1989 a survey by the UK newspaper *Computer Weekly* showed that recruitment adverts for language skills during 1988 demanded COBOL in 50.8% of cases, viz. 14 659 out of a total of 29 246. The 4GL demand lay third at 7.5%!)

A keen minority of end-users will of course be out there at the leading edge of computer use, but even they are very likely to be a little short of the 'system' aspects of the organisation's information management problems. And the remainder will almost certainly be seeking assistance from time to time, and on many of those occasions from their systems analyst colleagues, especially where the problem relates to fitting the applications they have developed on a personal computing basis into the integrated information systems of the corporate body.

Certainly also there *is* going to be a gradual but ultimately almost universal changeover to 4GL usage, or some equivalent. But woe betide the organisation that believes this is just a question of 'simpler' programming. The use of these more sophisticated code generators will demand considerable design ingenuity to ensure their use in an efficient and relevant way.

The systems analyst will, though, have to adapt to survive, there seems no question of that. And one of the firmest challenges in this respect is in the work related to knowledge processing. Expert Systems (ES) are inevitably going to play ever more crucial roles in the search for competitive efficiency, whether it be in the exigencies of the financial world or in the laboratised world of engineering design and development. During the 80s the development of ES has been in the hands of a few specialists, many of these, though not all by any means, systems analysis converts. This may prove to be a one-way process, with the ES developer never being expected to turn a hand to 'general' systems analysis work, but in any case there can be little doubt that the systems analyst turned ES specialist has a better chance of seeing the wider organisational perspective than the pure ES developer. This may have vital importance in relation to effective use of ES and even in such particular areas as systems safety—though that may prove to be yet another specialisation!

In the next section we shall be turning to the question of the human communication skills demanded of the complete systems analyst and enquiring about the part to be played by logic in that area.

1.2 The logical communicator

1.2.1 The importance of communication

The skills of human communication are of absolutely fundamental importance in systems analysis work. Although practice has sometimes sadly fallen short of the needs implied by this message it has always been heartily preached, even in the very early days of systems analysis. The shortcomings themselves carry important lessons for the future, especially as they have even greater relevance as data processing gives way to knowledge processing in many areas of application.

We can consider the problem under two main headings—the design of the human–computer interface (HCI) and the determination of the system's requirements. [HCI has also been known as the man–machine interface (MMI). The term 'interaction' is also a suitable substitute for 'interface'.]

1.2.2 Interface design

During the 80s interface design has blossomed and the benefits are being felt throughout the computer-using community, especially the community of personal computer users. It covers a wide range of considerations from the highly technical, such as design of voice input/output hardware, to the relatively abstruse, such as development of interaction through natural language (but notice the strong human connection between these range ends), while incorporating in the middle such practical issues as designing for user comfort (mental and physical) and system safety through effective man–system partnership.

Many of the sophisticated features of the interface will in future be part and parcel of the packaged software and the competitive edge may well be determined as much by the perceived quality of the interface as by any other factor. However there are bound to be many instances where systems are being designed at such a detailed level that the interface is an element of the design rather than a bought-in feature. Also the bringing together of existing applications under one network umbrella is likely to be an increasingly frequent systems design task, and the interface could well end up as a mish-mash unless specifically earmarked as a design objective.

Interface design is already the subject of complete volumes and it is not the intention to explore it widely here. But there is an important aspect of interface design which *is* our concern and that is the way in which both human and machine 'cognate', that is to say carry out thinking-type behaviour. It is probably not justifiable to talk of cognition in respect of the machine itself, but there is clearly an important element of cognate design necessary in order to present a machine to the user that enables that user to interact with it in an intellectually effective way. This is where the justification for looking at logic and knowledge representation comes in.

It must not be assumed that the interface can be designed to cope with the human contingencies in a purely logical fashion. Sure enough sometimes the system will be able to tell a designer that what is being attempted simply does not make acceptable logical sense—this is common enough in the syntax checking of submitted language statements—but there are equally situations where the logic is acceptable to the system but unacceptable in the context, and this implies that a deeper level of understanding is demanded of the designer. Beyond this logic of design there is the question of logic of performance—anticipating how the user will think, know and behave when faced with the interface under live conditions.

Of course a lot of this is down to knowing how to use the tools of design—the computer system plus the development software—and then using them with skill, common sense and imagination. But as the remit for systems analysis widens to include the installing of intelligent systems such as ES it is certainly not enough to rely upon flare and experience alone; the irrational and illogical still have little to offer the workplace systems analysis, even if there may be apparent room for them in home and arcade computer games.

1.2.3 Understanding user requirements

There have long been complaints that systems analysts time and again have failed to implement what the users actually need. There are many possible explanations of this to choose from; failure by systems analysts to ask the right things (poor investigative technique), failure to interpret the users' explanations, failure to offer the user the right alternatives, failure to recognise that users' needs change as their circumstances change and as they learn more about what is even possible, and finally failure to cope with users' propensity to get systems to fall down, either deliberately or by accident.

Many of these failures are down to poor training or poor application of training for whatever reason. But even with proper training and the best will in the world to do a good job there is still an aspect of requirements determination which is to do with knowledge about how to investigate in a logical way and how to model findings in a way that is then amenable to interpretation as clear system specifications. This is even more the case where applications are related to the skilled or 'expert' user and where the system to be designed is related more to knowledge processing than to data processing.

1.2.4 Cognitive science

Cognitive science has some relevance at this point, or at least it ought to. It is a relatively young science and definitions of it will differ in detail but we can say that it takes as its centre of focus the human mind (embodied

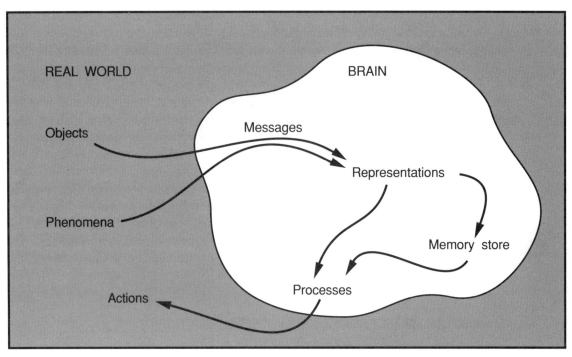

REAL WORLD

BRAIN

Objects

Messages

Representations

Phenomena

Memory store

Actions

Processes

Fig 1.1   The human brain as information processor

in a physical brain) and its function as a processor of what its owner finds in the world. It assumes we can accept that it receives and transmits messages (or data or information) which relate to the real world and its objects (any item of reality, not just physical objects), that it has capacity for storing and retrieving this information within its own compass, and furthermore that it can somehow transform information in such a way that the transmissions out into the world almost certainly have the capacity to change the world.

The question of how information relates to the world is very important indeed because it seems to be absolutely fundamental to accept that the mind uses fewer parcels of information than the number of objects it perceives; it re-uses the same ciphers or symbols and interprets according to context. The use of number is perhaps the most obvious of such re-use—'twenty' stands for twenty of anything your mind cares to think of as being countable. So it is essential that we know the context of the thought processes if we are to claim to understand another person's meaning.

Number illustrates another type of context too, that of a processing system. Once we have learned that system—how to count, in this case—we should understand the processing that is implicit in any symbol taken from it. The tables are turned here (homonyms instead of synonyms), in that we should understand twenty, 20 decimal, 10100 binary, XX Roman Numeral and so on, to be different symbols of the same intended meaning.

Not surprisingly, in reality confusion abounds and human communication can seem to be a very unreliable business. But this flexibility of languages is also its strength, as long as its users are ready to adapt according to context. Users of early programming languages will be only too well aware of the price of trying to use an inflexible language. That price is volume as much as anything—so much needs to be said in order to achieve the simplest process. However we shall return again and again to this question of interpretation of intended meaning, because however powerful the languages of logic that we devise its practical use is limited if there is an inadequacy with respect to shared meaning, whether it is the sharing of meaning between two human minds or between mind and a machine.

Let us consider for a moment an intriguing aspect of these mapping systems for representation of meaning. If we accept that the context is always potentially decipherable, even if the effort might be prohibitive in many actual cases, we might reasonably set as an expectation that if we freeze an information processing mind at any point of time we should be able to say what each current symbol means. But because we may be looking at a mind in the middle of a system process this might well be impossible. It is in the nature of symbol processing systems, even such familiar ones as number arithmetic, that they take on a life of their own with the effect that, until they arrive at some result or goal, meanings are pretty well totally obscure. Mathematicians are entirely at home with this situation and can happily process the most intricate and obscure-seeming symbol systems without worrying whether or not the actual meaning can be deciphered. Indeed they sometimes arrive at results, the meaning of which, or perhaps one should say the significant meaning as far as the rest of the world is concerned, is not at all clear to them. Thus it may be decades before some newly discovered mathematical system is found to be applicable to a practical problem-solving context.

This is not necessarily the case with all symbol systems of course. The systems we know as programming languages, for instance, have almost all been developed very much in tandem with their perceived applicability, and therefore in line with their sheer usefulness. So, while mathematics and mathematicians are usually given a wide berth (and even a bad press) by the non-mathematician, it is sometimes surprising to find the same non-mathematicians contentedly investing considerable time and effort into learning the symbol systems of some programming language or, more especially, of some software package. Not always though, for the bad press often rubs off onto computing so that it is still too common an assumption, for instance, that the mathematically gifted child is the obvious candidate for learning to use the computer. Children themselves know differently and it is not necessarily the budding mathematician that one discovers playing computer games. These children have learnt that the symbols do not have to be part of a mathematical system but can, as in fiction or in sports, represent human characters, personal decisions or tactical moves.

How is all this relevant to our theme here? Increasingly, computer software systems are being expected, indeed required, to become adjuncts to the processing systems of the human mind. Wordprocessors and spreadsheets are very good examples of how such ambitions are being modestly achieved at the personal scale of everyday office and home activity. The systems analyst is going to have to cope with this on a more general scale, integrating databases and knowledge processing systems into user interfaces designed for the mythical average user of the system. Understanding how to achieve effective knowledge representation and the processing of ideas is the order of the day here.

1.2.5 Face-to-face communications

The systems analyst who is unprepared psychologically or due to inadequate training to engage in direct communication with people is, in truth, no systems analyst at all. Facts can of course be unearthed through document searching, opinions sought through questionnaires, and systems designed and developed while closeted in an office seated at a desk or, increasingly more likely, at a computerised workstation. But avoidance of people—whether the managers, the technicians, the clerical workers or the operators—who are affected by implications of a new system or by modifications to an existing system is a pretty certain recipe for disaster in the long or short term.

How systems analysts should be made ready for this aspect of the work is not our concern here. We are concerned with how an understanding of logic and knowledge processing relates to such activity. First we must acknowledge that human communication is a matter of whole-person skill, where innate propensities are very influential but where good training can play a major role. It is not a question of being logical or even for that matter rational; indeed some features of a human-communication relationship can be spoiled by a single-minded attention to pure reason. But if we grant that the relationship is nurtured successfully by whatever human skills are appropriate we can confidently add that an ability to resort to logic and reason is essential. An inability to do so at crucial moments, such as being unable to explain for instance what is the logic behind the working of an interface if the would-be user asks about it, or being unable to model the user's knowledge when this needs to be captured by the design of a support system, could be well nigh disastrous. But the ability to call upon logic and well-structured argument at other times will also distinguish the truly competent systems analyst from the merely adequate.

A sometimes neglected skill of face-to-face communication is the apparently passive one of listening. The passivity is only apparent because expert listening involves comprehension, interpretation and preparation for follow-up clarification through questions and other types of prompt. Once again we must acknowledge the importance of personal skills in maintaining a sympathetic atmosphere which will enable the listening to continue, but there is here an added importance to the logic and reasoning in the mind of the listener. Even if illogicalities and faulty reasoning on the part of the speaker might need to be handled with diplomacy and, perhaps, set aside pro tem, it is essential that they be spotted at the earliest opportunity and earmarked for future investigative action.

Finally, there is the more formal type of communication expected of the systems analyst which involves the presentation of information to others in groups, small and large. This can range from straight information-providing talks to quite intricate technical training sessions. The preparation for such work is almost as important as the live session itself, especially where a series of structured learning periods is planned. The 'logic' here is rather more to do with the nature of the material to be dealt with, and certainly the ability to get arguments across is greatly enhanced by native and trained skills. But it is quite likely that explanations can be clarified dramatically with appropriate illustrative diagrams and tables. Since both are often used in this book to illustrate the same problem, there is a good chance that they will be seen to have value as presentation tools on a wider basis.

1.3 The rationale of data structures

1.3.1 Evolutionary processes

The systems analyst has to analyse information and data in rather special ways, having to keep in mind a number of different questions regarding how data will exist and behave once it is part of the basic elements of the complete information system—computer-based, of course. How will it be represented? How actually recorded physically? How coded and optimally reduced or enhanced? How will it need to interact with processes, either as the subject of enquiries or as trigger to other processes? Data is merely a representation of objects in the real world but the allocation of that representation has to be according to a very clear system since, once within the boundary of the computer system, there are strictly limited opportunities for adaptive and intelligent interpretation such, as we remarked, occurs in the cognitive processes of the human mind.

Early computer-based information systems achieved the clarity of representation by being bound by extremely stringent rules of development. First, the types of data considered suitable for computerisation were constrained to being either the numeric or the straightforward labelling types. Book-keeping type accounting systems fitted this constraint very well and this therefore constituted a very large proportion of candidate systems.

Second, the programming languages were the equivalent of coded step-by-step instructions, thus making all data processes and data movements painfully visible, in the sense that data formats needed to be explicit at all times. This consequently demanded of the data a very rigid format. It was almost as if every character of every piece of data had to be manually pigeonholed.

Third, there was the problem of the very literal filing system offered by, for example, magnetic tape reels. It was true that data could be passed back and forth in bundles or *blocks* of characters but these blocks had to be very strictly defined and formatted or even more horrendous complexity would be demanded of the programs using them.

Fourth, when there was any assistance from the computer system itself in the shape of an operating system or other programmer-aiding software pre-written as re-usable chunks of coding, the rules of usage were special to that computer, that operating system and that set of programs. One step out of line and the programmer/analyst would find painstaking work riddled with bugs.

Things had to change if systems development was not to remain a slow and esoteric (and very uneconomic) activity. The first constraint has been loosened in a number of ways. The database approach to data storage encourages the designer to think in terms of variability of record format, though the general type of format is still quite formalised in most database systems to being a stylised arrangement of item types within a record type. Word and text processing has become commonplace as part of the personal computing sphere of activity. Graphics applications are becoming increasingly common, ranging from the so-called business graphics attached to spreadsheet packages to the art graphics of screen painting packages and animation systems. Current developments are starting to bring all these representational forms together into encyclopaedic pools of knowledge. At the same time, knowledge processing ideas are freeing the designer from having to think of all this stored information as separate little pieces of data.

The second constraint of the primitiveness of available programming languages has again been loosened by the development of higher generation languages, though there is still a lack of proven experience with these, with the associated problem of lack of skilled programmers at this level. It is in the field of personal computing that again we find more dramatic loosening of the stays—integrated packages provide the enterprising user with powerful data manipulation tools which require some, but not impossibly high, investment of effort in learning the interface languages.

Filing systems, our third constraint, have developed almost beyond recognition. Storage capacity is hardly ever technically limited—though cost may inhibit still—and the arrangement of data on the physical media is handled by sophisticated intermediary software which hides most of the detailed and technical problems. The systems analyst is still well advised to try to understand what is in fact going on in order to use these software systems efficiently, especially where large volumes of data are concerned, but it is more the case of finding out what can be done with what is available rather than of worrying about how one is constrained.

The final constraint, that of the operating environment, although there have been enormous developments, is still perhaps the most restricting. Standardisation of operating systems has some way to go before the system developer can feel relatively free of restrictive practices. The rapid growth of networking has meant the exacerbation of the problem because of the general moving of the goalposts, with, at the same time, the hastening of standardisation because of organisations finding themselves with the practical problems of systems (hardware and software) that refuse to communicate with each other.

1.3.2 The development of data modelling

These evolutionary processes have not however been simply a matter of presenting the designer with improving quality of material. There has had to be a matching development in the ways in which the designer thinks about the problem of representation. While aware of having to work within very tight constraints the designer's thinking was limited to fitting simple data designs into the simple formats available—a tedious though relatively non-complex task, especially since the designer was able to get away with telling the mystified user that the data had to be this way because the computer required it so. But with freedom from constraints comes freedom of design choice and the more challenging task of finding out what the user wants and needs in order to be able to make best use of information; and then using the by now much more sophisticated software tools to see if what are bound to be more complex arrangements can be specified.

This description is beginning to sound more like design in an engineering environment. Engineers find that they need very often to resort to some kind of modelling if they are to be able to communicate with clients about the options and possibilities. Indeed they more often than not need models in order to develop and test their own design ideas. And this is what has happened in systems analysis. In order to understand user requirements, to talk over the possibilities and in order to think through the technical design options and feasibilities, the systems analyst too has to build models—data models

Data modelling was given its initial impetus as a support to making effective use of *database management systems*—software packages developed as tools for building data structures with adaptable formats. Modelling took on the form of box-and-line diagramming with a limited number of conventions which enabled the designer to see the connectivities between what had, prior to this, always been thought of as separate files. Once learned, these diagramming techniques could be used on a pragmatic basis, sketching as it were the skeleton of the possible structures. The question of the logic of the design hardly arose at this stage.

When some of the disadvantages of this approach began to emerge—the rigidity of the design once implemented, for one—the thus far rather neglected ideas of mathematical relational principles due to Codd began to be given more notice. While being more difficult to grasp at theory level, the relational approach offers designed systems which are much more easy for the user to handle through reasonably straightforward access or enquiry (or, simply, query) languages. Understanding the relational approach actually gives one insight, retrospectively as it were, into the earlier pragmatic modelling methods, and nowadays the two approaches (a range of approaches in fact) sit together reasonably happily.

So that brings us, once again, to our justification for learning the underlying logical principles. The systems analyst could be limited to seat-of-pants designing if this understanding is lacking. Pragmatic principles of design are undoubtedly an important part of the designer's tool-box, but an understanding of the logic theory gives an important extra tool—that of being able to switch from one technique of modelling to another, from one database management system to another, from one query language to another, without either getting totally lost or having to learn the different grammars rote fashion rather than from a basis of understanding foundation principles.

As a final note to all this design history, it is worth asking whether it wouldn't be better to have a single modelling method, database system and query language. The simple answer to that is: "yes, of course". But life is never simple. While it is true that the variety has developed out of expediency and experiment, there is no guarantee that we are yet ready for the definitive version of any of these. For example, SQL is often these days referred to as the de facto standard query language—and we shall give it due consideration in this book—but ask any computer language theorist how suitable it is for this pinnacle role and one will no doubt get some cynical replies. Therefore we should not be surprised if there continues to be a modest Babel of query languages actually in use for some time even though SQL may well be market leader and it should help to remind us continually that the search for a single standard is a relevant issue. The systems analyst with a basic understanding of the underlying theory will be capable of resisting despair while the Babel stays with us.

1.3.3 Data, information and knowledge

If the argument of the 70s about what makes data become information confused the average practising and practical systems analyst (we have already suggested that information can be thought of as data which has come within the range of some human's perception), then the argument of the 80s about what turns information into knowledge has probably confused ten times more. And yet, knowledge representation, knowledge processing, knowledge engineering and, even, knowledge workers became phrases of pretty common currency in the late 80s. Whether all the users of the terms agree on their meaning, or even have a clear meaning in their own minds, is debatable. We hope this book will help the average systems analyst to understand and use such terms with greater confidence.

We start with a deliberately blunt question. Isn't knowledge just another way of saying data? Data which is storable, processable and retrievable by the human mind, but just data nevertheless. The answer is yes, but the criteria for being storable, processable, etc, is what actually turns data into knowledge!

Let us take a relatively straightforward example. Suppose you are a Sales Manager for a medium-sized toy manufacturer and an assistant one morning drops on your desk your requested data on how sales moved last year. It is set out in the sequence by which the assistant managed to obtain it. From one sheet to the next there is no telling what piece of the facts is going to turn up (see Fig 1.2). The facts are all there, but they are quite

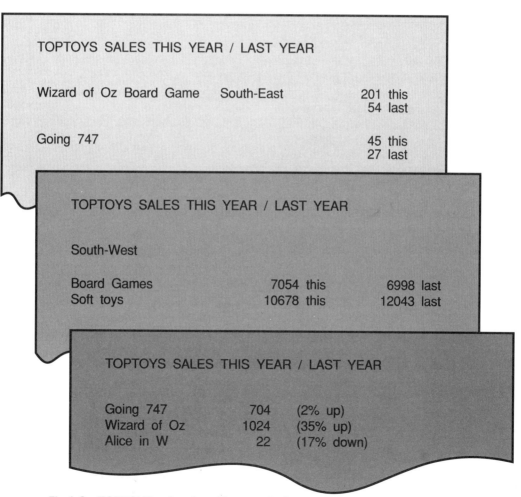

Fig 1.2 TOPTOYS sales data (first version)

useless in relation to your function as decision-making Sales Manager: "I have the data, but they give me no information."

You pass the data to a senior assistant with instructions to put them into the form of a statistical report—a report showing how each toy type sold by region and by outlet, showing how each region performed for all toy types, how each outlet performed, what the profiles of salesman-outlet pairings are in terms of what the outlets ordered (rather than what they actually sold), and for each toy type a month-by-month profile of total sales (see Fig 1.3). When this report drops on your desk a few flicks of the pages and you begin to feel the warm glow of becoming well informed about what happened to sales last year: "Thanks, I've got the information I wanted".

You start to digest this information and soon some serious questions start to occur to you. Your Going 747, the large-scale airliner, seems to have been very popular all over the place last year, but will that be repeated next year?

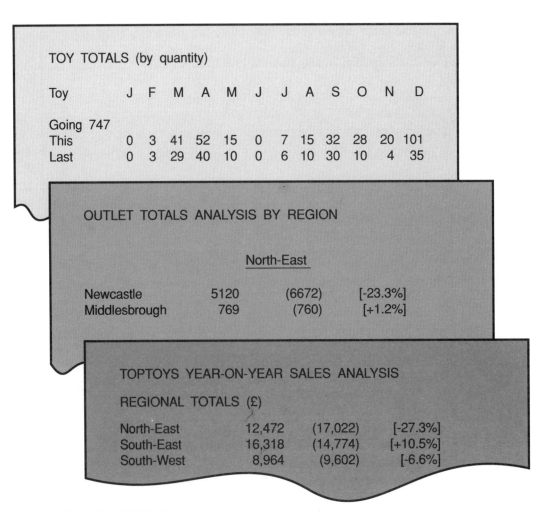

TOY TOTALS (by quantity)

Toy	J	F	M	A	M	J	J	A	S	O	N	D
Going 747												
This	0	3	41	52	15	0	7	15	32	28	20	101
Last	0	3	29	40	10	0	6	10	30	10	4	35

OUTLET TOTALS ANALYSIS BY REGION

North-East

Newcastle	5120	(6672)	[-23.3%]
Middlesbrough	769	(760)	[+1.2%]

TOPTOYS YEAR-ON-YEAR SALES ANALYSIS

REGIONAL TOTALS (£)

North-East	12,472	(17,022)	[-27.3%]
South-East	16,318	(14,774)	[+10.5%]
South-West	8,964	(9,602)	[-6.6%]

Fig 1.3 TOPTOYS sales data (revised version)

What do I know, you ask yourself, about current events, changes in fads, influences of popular entertainment or plain old-fashioned sales cycles observed so often in the past, that will have a possible bearing on future sales of the Going 747? Similarly with the Alice in Wonderland Board Game, which fared so disappointingly last year. Are there any plans for the re-release of the Disney full-length feature and if so what effect can we expect from that? An intensive all-night brain-storming session with your trusty colleague, the Marketing Director, is called for: "I've got here the statistics on last year's sales; gives us all the information. But what do we really know about what it all means in terms of this year's schedules? I'd like us to get together and think that through."

Any systems analyst should have been entirely happy with the earlier situation of the request for a statistical report. Indeed it is almost certain that a system could be set up which automatically captures the relevant data as it becomes available, stores it and partially processes it into skeleton reports, and is not only ready to churn out said reports at any time but is also accessible through a terminal for responding to requests for format changes to the reports or presenting the reports as histograms, graphs, pie charts or whatever. But what about the later development in managerial thinking? Is this in the province of computer-based systems or is such knowledge buried deeply and somewhat inscrutably in the minds of the two managers? Some of it seems even to be beyond their processable knowledge unless they have powers of clairvoyance!

It should be evident that what is referred to as being knowable in the final discussion covers a range of *levels* of information. Knowledge of current events, for example, may seem like factual knowledge, but it is a good deal more than that since it is not only the fact of the current event but knowing that it is germane to the sale of the Going toys that is relevant. This is classified knowledge in the sense that stored data needs explicitly to belong to appropriate classes, possibly set within a hierarchy of such classes, for it to be retrievable under any future call for it. This is very different from being stored in a labelled file along with a family of very similar facts. This is only slightly more sophisticated than the idea of information in the statistical report, but it is different by virtue of being information with potential relevance in a host of circumstances rather than simply in the context of the statistical report only.

Knowledge of changes in fads is different again since it brings us squarely into the province of subjective information. A fad to a statistical analyst is recognisable by its distinct 'peaky' profile, to a sociologist by its evidence of influencing behaviour, and to the production planner as being a nightmare if you fail to anticipate it. A systems analyst may be able to cope with incorporating this knowledge as information if designing a specific system for a specific user—for example, building into the program a peak-recogniser for the statistical analyst, or a time-series forecasting algorithm for the production planner.

The sociologist's view is more difficult to cater for since there are probably different ways of perceiving a fad even within a close-knit team of sociologists. Perhaps here then it would be necessary to develop a special information analyser geared to the type—or 'style' is maybe better here—of analytical activity that sociologists would engage in.

Knowing about possible future events may not be a matter of clairvoyance but to do with the comprehensiveness of the data that is scanned. How can you afford to have stored in your system all the future schedules, as far as they go, of TV stations and film distributors? If you *can* afford access to some appropriate proprietorial database, how can you efficiently filter through it all to yield the facts that are relevant?

And, assuming it to be a feasible scenario at all, after you have negotiated all these problems, what then? The experienced sales manager will recall that TV programmes can influence sales in a certain way and, if pressed, might even be able to express it as some kind of rule, such as: "expect the re-release of the Disney Alice to improve sales between 10 and 50%". That may sound a very wide and therefore crude forecast band, but if that is what is successfully used in practice then it is the appropriate rule. That is real expert knowledge.

Now improving sales by anything in excess of 10% may require the triggering of some other action—here perhaps the placing of a discussion item on the next meeting of the Board. This is a definite cause-effect rule and ought not to rely purely on the memory efficiency of the Sales Manager, or whoever.

These last two examples—of rules for attention and for action—are not often to be found programmed into traditional data processing systems, partly because few people ever expected them to be, but mostly because it requires quite a bit of programming effort with conventional languages, and that's for just two rules out of what may prove to be a population of many thousands or more in the active day-to-day business. Indeed we are here looking at rudimentary knowledge processing as the term is most commonly used and understood. To avoid the heavy cost of programming such knowledge, special software tools have been developed. But these tools require the user to have some idea of how the basic elements of knowledge should be represented. And the whole business of getting the rules analysed, designed, programmed and into action is an important part of knowledge engineering.

So what has changed from what we first considered to be simply data and then later information? What are the characteristics of this so-called knowledge? At the crudest level we might just point to the result—that a fact of procedure, a way of doing or thinking, has been stored rather than the storage of just a bald fact (data) or a set of ordered facts (information). But these rules usually have further important features. A rule itself can be parsed into the sentence structure 'if...then...' and this can be seen as part of a hierarchy of such sentences (nested ifs) as any programmer will readily recognise. Rules can trigger other rules and thus act as parents to whole families of more detailed rules. Our example of "expect the re-release of the Disney Alice to improve sales between 10 and 50%" might simply be one instance of a more general rule that can be applied to a number of products with different names, different pre-conditions and different actions (eg. simply different percentage forecasts, in this case).

This might strike you as a very limited interpretation of what we usually think of as human knowledge. There are many refinements to be added to this if...then... skeleton, as we shall see, but it is true that it is far from being a perfect modelling tool for intelligent processes. What about the knowledge of physical skills for example, or the all-too-familiar 'gut' knowledge associated with experience and intuition? And what has happened to the question of meaning, the control of the semantics? Cognitive scientists are actively interested in extending the realm of artificial intelligence to encompass such knowledge types and issues, and indeed many other things. But we must set aside this question for later, since the early chapters of the book will be content with getting a sound grip on the simpler models of logic and knowledge processes.

It is worth concluding, though, with a thought about what lies beyond this scale of: data ... information ... knowledge. We have already used the word 'intelligence' and it is this which is probably the next marker on the scale—intelligence and learning, to be more exact. Indeed learning, in the sense of the process of acquiring the ability to acquire skills which in turn lead to further knowledge, is very much part of the systems analyst's remit, since designed systems should always be delivered with accompanying means of learning how to use them. The training of one human by another through face-to-face instruction or via system manuals is the traditional method, but demands of volume and efficiency are increasing the pressure to provide the means as part of the software—computer-based training (CBT) and computer-aided learning (CAL). It is certainly expected of the highly competitive packaged systems offered to personal computer users. It is interesting to note that this has brought us back once again to the question of the importance of understanding human communication.

1.4 What's under the bonnet?

1.4.1 The systems analyst as engineer

Much of what we have been discussing so far as being of importance to the repertoire of the systems analyst has been strongly related to analytical work and skills. But we must remind ourselves that the systems analyst is also a designer, and very often is also expected to do some of the building associated with the designs. In this aspect of the role we see the systems analyst as engineer, knowing not just how to use tools but having besides some understanding of how those tools actually work and moreover how they work with different materials. The tools are multi-layered—from hardware right through to software—and the materials range from the very tangible such as written program code to the rather intangible such as the prospective behaviour of users.

It is always a contentious issue in any sphere of human activity in the twentieth century as to how much the user of technology needs to understand about that technology. Do we really need to know what goes on 'under the bonnet'?

It has always been the author's own viewpoint that when the question is asked about the systems analyst the answer is an emphatic 'yes' with an immediate qualifying 'although'. 'Yes' because there is potential for improved efficiency in a system designed with technical understanding rather than blind use of what the tools offer, and 'although' because we need to show caution against becoming obsessed with technical detail, technical detail for its own sake. This is a difficult balance to maintain. In fact one problem is that the balance will need to be different for different individuals and for the same individual in different circumstances.

1.4.2 What technical knowledge is relevant?

Potentially there are many relevant areas: the area of fundamental principles of the logic circuits of the digital computer, based firmly on simple logic; the logical syntax encountered in almost every high-level programming language and absolutely crucial in some of the more recent AI-associated ones; the logic of relational operations as they occur in database systems and query languages, already touched upon under the issue of data modelling; the principles of operation of the inference engine, software that drives the knowledge processing in AI and Expert Systems.

The systems analyst could get by with next to no knowledge of what is 'under the bonnet' of these—indeed a great number are doing so—but it is our contention that this is not satisfactory and certainly not efficient. The tools are getting more sophisticated and numerous and it is getting increasingly less likely that the systems analyst can be a knowledgeable user of anything approaching all of them. More and more the need is to get to grips with the underlying principles so that we can migrate from one to another without starting all over again. This book, it is hoped, helps the systems analyst, and other readers, to be able to achieve this.

2 Two-valued Logics —a Simple Start

2.1 The story so far

A fair proportion of systems analysts will have met two-valued logic in some practical context or other. Programming languages, database query systems, spreadsheet packages and text-searching aids have for many years incorporated the logical functional operators AND, OR and NOT, together with the returned values from these and other functions of the logical results: TRUE and FALSE. These results will have been used in constructions such as

IF {X is TRUE} THEN {Do this action}

Many of the users of such facilities however seem to have learned to use them through practical experience—trial and error—rather than through an in-depth appreciation of what underpins their design. Others have remained very wary of their use because of apprehension that they are too mathematically oriented for the likes of themselves. The advent of knowledge processing, sometimes in very user-friendly expert system shells (software buildings aids), has brought home the unsatisfactory nature of this state of affairs, although many people, not just systems analysts, will still learn to be adept in the use of these by sheer practice.

Before we set about providing the opportunity to overcome these grey areas of knowledge it is worth noting that recent years have found the theorists and researchers in Artifical Intelligence (AI) and Cognitive Science increasingly dissatisfied with the crude limitations of the two-valued system of logic, ie. TRUE and FALSE based systems. We should not let this disguise the fact though that two-valued systems have enjoyed an enormous practical success and it is partly due to the prominence of this success that attention is drawn to their limitations and the need to develop things from here. We shall look at what these developments are, or need to be, mostly in the latter part of the book, but also occasionally, where it is thought appropriate, as we proceed. Initially we concentrate on seeking a real understanding of two-valued systems.

What two values? What system? The most familiar of the two-value pairs is, as we have noted above, TRUE and FALSE. But perhaps a more useful start is to free ourselves of that linguistic constraint. Any value-pair which has the style of a 'black' and a 'white' will do. Indeed in many contexts it is far better to think in terms of other value-pairs. They may be YES/NO (eg. in decision-taking contexts), PRESENT/ABSENT or MEMBER/NOT-A-MEMBER (eg. as applies to data items in records, or to records in files, etc.), ON/OFF (eg. as in program flag or switch settings), or one/zero as in binary arithmetic, the operational mode of digital computers. Indeed BLACK/WHITE itself may be suitable, as in selecting the dots of a newsprint-type picture.

As to the 'system' aspect we should say most emphatically that for one thing it stresses this: that, with whatever value-pairs we are using, the system of working with them has a consistency of operation that enables any one mode to emulate any other—provided, of course, that the *translation* (or *mapping*, as mathematicians would say) between them is clearly defined.

This principle lies at the heart of computing and, more especially, of knowledge processing. A most important case of such mapping is naturally that between the real world and any computer-based model of it. And therein lie the many arguments of the inadequacy of the two-valued system. In practice there is rarely a direct mapping between the real world and the computer-based model; standing at intermediary points will usually be a number of stages of translation and many of these will be the responsibility of the systems analyst. The early days of computer-based systems development were concerned with the machine end of the scale—the binary bits and program flags which were needed to get a program de-bugged. Nowadays much of that is taken care of by the software and it is the mapping of real-world meanings into the sentential logic (ie. modelling logic by means of sentences or statements) that is more the focus of attention.

Which brings us precisely to the question of the nature of our system. We find that the language of language-analysis (linguistics) provides us with a useful vocabulary here. First of all we need perceptible objects to work with—an **alphabet** of symbols. We need acceptable, meaningful assemblies and structures of these symbols—a **vocabulary** of words, phrases, expressions. We need rules of combination and order for these words— a **syntax** and **grammar**. (There is no need to worry about any distinction between these two words here.) And what we have then is a language. But because at every stage of creating that language artificially we will have been precise, consistent and clear (we hope!), we will have not a 'natural language' but a 'well-formed language' or, more simply, a **formal language**. This is what the wff (wuff) jargon of the logician/linguistician is about; it stands for **well-formed formula**, meaning that when we look at a string of symbols in its appropriate context we can rationally analyse why the symbols are where they are by using the grammar-scalpels of the formal language that generated it.

Programming languages, the languages of menu-systems with software packages and database query languages should all be well-formed but they are never confined to the purely well-formed rules of the two-valued universally-applicable language known as Boolean Algebra (but also by other names such as binary algebra, first-order logic, propositional logic, etc.). The name Boolean Algebra is used in recognition of the 19th century English philosopher-mathematician, George Boole (1815–1854), who developed a symbolic method of study of what he called 'relations between classes'. It is also an historical fact that many version of, and developments of, this algebra have emanated from a number of other mathematicians, pure and applied, with the unfortunate consequence that a plethora of symbol systems (alphabets) have emerged. Sometimes this is useful in the sense that a given alphabet particularly suits the purpose to which the algebra is being put (eg. the keyboard-set limitations in a query language) but mostly it is an annoying obfuscation of meaning—never after all a welcome feature of a language!

This Babel-like variety of symbol systems runs throughout the world of textbooks and also nowadays through various programming and package-interface languages. If this is aggravating to the readers of mathematically oriented texts and users of multiple language and package environments, it ranges from the daunting to the devastating for the systems analysts or other generalists who are relatively uninitiated in the practised use of symbol systems. At least, systems analysts will be better prepared psychologically for this lack of standardisation, being only too familiar with the problem in the competitive world of software manufacture and marketing, where variety of interface is often related to matters proprietorial. But it is nevertheless a problem and a problem that must be faced. We shall try to show the extent of the problem shortly and suggest that there is a virtue in having some variety, while keeping the variety used in this book down to a minimally optimum level.

Although it is possible for the mathematically inclined to 'play games' within the boundaries of formal systems—and this provides very useful, sometimes essential, discoveries of operational improvements for them—the systems analyst is concerned almost entirely with the mappings themselves; the mapping FROM the real-world events at one end of the range right through to the mapping TO the working computer-based systems at the other. Bearing that in mind we shall try always to keep explanations and examples relevant and meaningful.

2.2 Sample two-valued systems

2.2.1 Physical switches

Let us consider a system comprising connected switches which allow or prevent a flow of electrical current through the connected circuit. It is a true two-valued system if each switch is positively either ON or OFF; no in-between 'leakages' can occur with these switches. We will allow only one input line to the whole circuit and one output line, thus enabling us to determine whether the circuit as a whole is ON or OFF. (It could itself act as a switch in a wider circuit therefore, a very important principle of logic systems as we shall see.) By physical observation, or indeed by reasonable and intelligent use of imagination, we should be able to see that we can formulate the following rule within such a system:

If a circuit comprises a series of switches connected output-to-input (see Fig 2.1(i)), the circuit can only be ON if each and every switch is ON; a single OFF switch will be sufficient to switch the whole circuit OFF.

We might arrive at this conclusion by observing it to be true for one switch

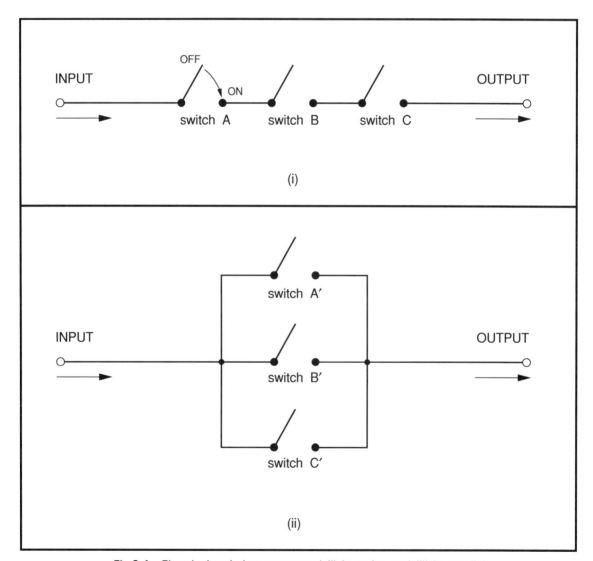

Fig 2.1   Electrical switches connected (i) in series and (ii) in parallel

(trivial), two switches in series, three switches in series, and so on. The generalisation to any number is interesting and an example of the power of discovering rules within any system.

Now look at Fig 2.1(ii)). Here the switches are connected in parallel and it should be possible to see that we can formulate another rule for this arrangement:

> If a circuit comprises a set of switches, each of which has its input connected to the main input and its output connected to the main output, then one switch alone being ON is sufficient to switch the whole circuit ON.

Computer logic gates are designed on the principle of switches, albeit at the molecular scale of materials that allow or inhibit the passage of quantities of electrons. These gates are then switch circuits in their own right: computer logic circuits made up by connecting together gates are more sophisticated in principle and we shall return to these shortly.

Electrical and electronic switching fits our two-valued system constraints very well. But with other types of physical circuit there are problems. Liquid- or gas-flows in a valve-controlled circuit, even supposing the valves can be, in effect, perfectly ON/OFF tight, take time to settle to a steady state so that after one valve in a series is turned OFF there may be a continued pressure in other parts of the circuit over some noticeable time lag. However the steady state will again obey the rule we observed above. Actually the same problem occurs on a very fine time scale with electronic circuits, which is why a computer needs some kind of clock mechanism to define each steady-state micro-period of time. It is an interesting example of successful mapping between the real world and an ideal model!

2.2.2 Decision taking

Switches in either ON or OFF states are one thing. Decision-taking systems where each decision can result only in either YES or NO are probably rather harder to swallow because we tend to think of decisions as meaning 'human decisions'. But the setting of a switch is a decision and so we can take this as an example of the simplest mapping of decisions. The decision to have a switch set to ON is here mapped on to the reality of its actually being set to ON.

In any case we can easily imagine a system of connected *binary decisions* (questions admitting only yes/no answers) serving as a model for taking the decision maker through a complex decision situation step-by-step. This occurs with aids to the untrained in deciding what to do (eg. equipment fault-finding), what classification their problem situation has (eg. potential benefit claimants), or a specification of proper procedure (eg. clerical document handling).

If we try to parallel the discovered rules in Section 2.2.1 we find it can only be identified with a string of decisions connected by their YES outcomes for case (i) and their NO outcomes for case (ii) (see Fig 2.2). Any other answer is sufficient to remove the decision-taker from the path. This feels less universal than before—yet it is the same principle in a different guise. We shall explore this further in Chapter 6.

Decision modelling is certainly a much more complex subject than is implied by these simple examples, but for the moment we are sticking to our objective of making a simple start. The more complex issues can be discussed later in the light of what is hoped will be a better understanding of the fundamental principles.

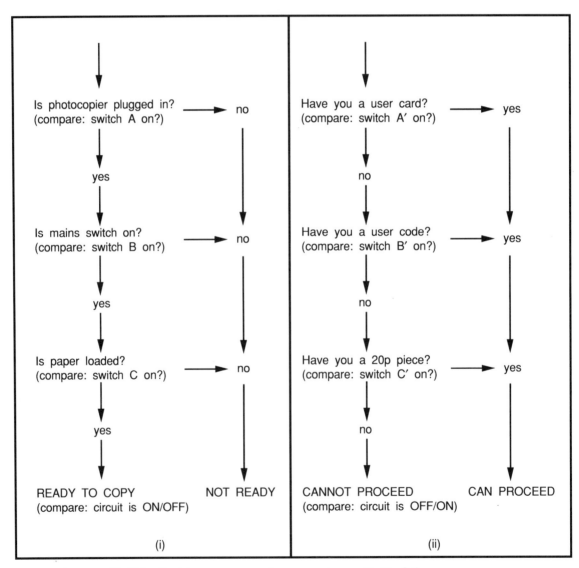

Fig 2.2 Decision sequences for comparison with Fig 2.1

2.2.3 Class membership

There is an immediately obvious connection between a system of allocating (or analysing) memberships of categories or classes (or sets) and the taking of binary decisions as discussed in Section 2.2.2 above. Membership is, at its simplest, a two-value system because we are asking: is such-and-such a member of a given class, and expecting the answer YES or NO in return. Only if the answer is YES to a number of such questions will the subject be classified as belonging to the fully categorised set as equivalent to our case (i) and the answer has to be YES for only one question in case (ii) (see Fig 2.3).

But here the 'feel' of the problem is different again. For one thing it would appear that the sequence for asking the questions is immaterial. And yet our observed rule is still there: any one not-a-member case is sufficient to disqualify membership from the final categorised set. (This is like saying that the sequence of connection of the switches in Section 2.2.1 does not affect the rule. However if the switches were also part of other interconnected circuits then re-arranging their sequence would affect things materially.) Another interesting angle is that any subject-problem item belongs to a potentially large number of categories at the same time. In our example the subject belongs, for instance, to each of the categories which has the single feature indicated by answering each one of the questions in the affirmative, or the categories which have feature-pairs and feature-triples and so on. (For those already familiar with relational databases this is like the situation where every record is a potential member of a large number of potential table-relations.)

2.2.4 Propositional statements

A propositional statement in a two-valued system can be thought of as the binary decision with the interrogative (the question mark) removed. The affirmative statement must therefore be either TRUE or FALSE. This is a somewhat ruthless constraint if we compare with the general statements of natural language in written and spoken communications, and even more so if we compare with thinking processes. In actual communication we have a range of features which mean that every statement made is open to question—about the intended meanings affected by emphasis, context, accompanying gesture and the status of the writer/speaker, about what period the statement is meant to apply over, about how judgemental the statement is either deliberately or inevitably, and so on. Herein lies a good deal of the inherent dissatisfaction with the real effectiveness of two-valued systems for AI purposes.

However, again we must make our simple start and study propositional statements as if they did truly fit the constraint. It is a steadying thought to consider the point already made—that the systems analyst has to face the

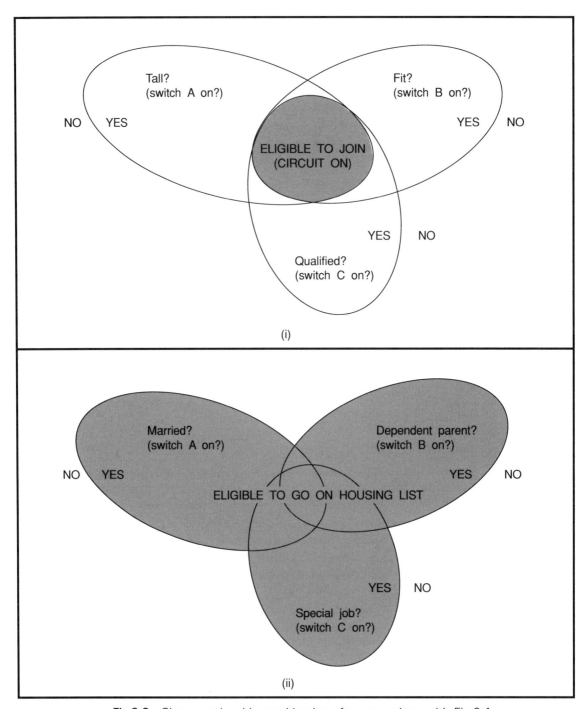

Fig 2.3 Class membership combinations for comparison with Fig 2.1

The compound propositional statement:

THE COMPANY PAYS ALL EMPLOYEES REGULARLY (TRUE)
AND ALL EMPLOYEES HAVE A BANK ACCOUNT (TRUE)
AND TRANSFERS ARE FULLY ELECTRONIC (FALSE).

is FALSE because one of the component propositions is FALSE.

Fig 2.4 How a compound statement is made FALSE by containing one FALSE component

problem squarely because designs are in effect the mapping of awkward and capricious human communications behaviour on to the strictly binary world of digital computing machines.

If we look again at our two cases in the context of such statements we find another unnatural-seeming feature. This is that, in a compound statement made up of a collection of elementary propositions, the rule would, for case (i), mean that it will be falsified by any single one of the elementary propositions being FALSE. Look at Fig 2.4. It can seem unreasonable that the whole statement should be FALSE when so much of it is palpably TRUE. This is where logic starts to feel very different from everyday communication in another more crucial area, that of so-called reasoning. We appear to reason, to present an argument or a case for instance, in anything but a logical way unless we happen to be a mathematician engaged in a strict proof or, perhaps, a scientist propounding a theory (though often enough such expositions are riddled with difficulties of meaning and belief).

And yet we are increasingly attempting to use such strictly logical tools in applied computing since Expert Systems Shells offer to try to emulate our everyday reasoning. Put in this way, things look bad. However, they are not as bad as they seem, since so much depends upon how we make use of our tools, recognising their limitations as much as what they offer us in facility and function.

2.3   The common connectives: AND, OR, NOT

2.3.1   What connectives are

Let us return to our linguistics terminology of Section 2.1. We have so far said little more about our two-valued systems than that they have somehow to embody this binary-value feature. But what is it that has to be binary-valued when we compare with linguistic terms? A switch, a decision, a membership element, a proposition? What generic description can we give all these types of two-valued things? The mathematician's answer is **variable**, the noun not the adjective. A Boolean variable is an entity which can take one of only two values. How we depict this variable will depend upon our alphabet of symbols, which we have not yet defined. Since all we need for a variable is to be able to label it and recognise it, we can use pretty well any word or phrase we like; whatever is convenient in context. 'Yellow switch', 'Is it yellow?', 'All yellow switches' or 'The switch is yellow', are all possible labels under the different types. On the other hand we could simply use the character y, as in common algebra, so long as we can see in context what y stands for.

So somewhere in any presentation of logical reasoning using a formal system, we expect to find a definition of any of the labels used.

But we also know that we shall want to 'do things' with our variables. The examples running through Section 2.2 started out as rules concerning connected switches. The common idea in all the types of system is that we wanted to relate or connect variables in a particular way and then say something about the behaviour of that relationship or connection. We need to introduce special words into our vocabulary that describe these relations or connections. We shall call them **connectives**, though other words are also used, eg. we referred to them earlier as logical operators. In order to take any mystique out of this jargon, think of the plus and minus signs, etc., in arithmetic as connectives for specifying operations between numerical values. We do need sometimes to make it clear which system we are working with by using appropriate descriptive prefixes, eg. *arithmetic* connectives, *Boolean* connectives (or logical connectives).

While we are about the business of defining terms it is an appropriate point at which to mention the problem of language as language. With natural languages we generally use language itself to discuss the language, though the grammarian will use special words to do this: verb, noun, pronoun, noun-phrase and so on. When we use one language to describe and, in formal systems, to define another we need to be conscious that we are using it as what is called a *meta-language*. For the purpose of setting up formal systems:

> We cannot accept the usual vagueness and ambiguity of ordinary language ... . Formalisation requires rules and the rules which make a system precise are themselves relatively informal, therefore we have to think in terms of a hierarchy of languages. [George, 1977]

The Russian-doll nature of this problem should be taken very seriously. We shall appeal to thoughts about this problem often as we proceed. The systems analyst is responsible for the quality of meta-languages at the top reaches of the hierarchy (eg. the design specifications), and illogicalities, poor grammar and misuse in these will permeate the languages of the lower reaches of the hierarchy, ie. the implemented system!

2.3.2 The two-value pairs

We have deliberately been using a number of different pairs of terms for Boolean values up to now: ON/OFF, YES/NO, MEMBER/NOT-A-MEMBER, TRUE/FALSE. This has been done to illustrate from the outset the variety of interpretations that are possible of what two-valued systems are like. For the remainder of this section we shall use just TRUE/FALSE in order to introduce the common connectives in a consistent fashion. Later we shall return to other pair-values in order to enrich the understanding of logical processes. For the moment it is worth thinking about the equivalence in the ways shown in Fig 2.5. This may look rather clumsy but it should help in the interpretations of TRUE/FALSE in following sections.

TRUE
It is TRUE that the switch is ON
YES; the answer to the decision question is TRUE
It is TRUE that this item is a MEMBER
The proposition is TRUE

FALSE
The switch is OFF; it is FALSE to say it is ON
NO; it is FALSE to say the answer to the decision question is TRUE
NON-MEMBER; it is FALSE to say the item is a MEMBER
The proposition is FALSE

Fig 2.5 Equivalent meanings for each of TRUE and FALSE

2.3.3 The AND connective

We have already identified something of the character of this connective, though without showing how it is actually used. In fact there are many ways of symbolising and using the common connectives as we shall discuss in Section 2.4. For the time being we shall use the word AND to symbolise the connective, and lower case individual letters of the alphabet to denote those things to be connected (ie. the labels for the switches or decisions or classes or propositions). We could represent our rule then as follows:

The expression {a AND b AND c ... AND n} is TRUE only if all of a, b, c ... n are each of them TRUE.

31

Alternatively, we could put it this way:

> The expression {a AND b AND c ... AND n} will be FALSE if any one or more of a, b, c ... n is FALSE.

Because the system is two-valued the 'is TRUE only if' implies 'otherwise it is FALSE', and the 'will be FALSE' implies 'otherwise it will be TRUE'. Also, note the use of curly brackets around a compound expression. We use this device to help make clear the boundaries of the expression. The expression itself is a Boolean or two-valued variable and could, if necessary, be represented by a single-character label.

We must at this point issue the sternest warning about not confusing AND with 'and'. (The statement "AND and 'and' are different!" illustrates very well what was said about meta-languages. Think about it.) The conjunction 'and' in natural language is used in many guises, but the one to beware most of all is its use as a kind of connector across time. For example: "The systems analyst gave the specification to the programmer and she coded the program" says rather more than the sum of the two parts. There is more than a hint of event-dependency, that the programmer could not have coded without first having the systems analyst give her the specification. It is always best to 'read' the connective AND as meaning 'and at the same time (literally and metaphorically)'. The time-dependency element can then if necessary be made part of the variable's label. For instance, "The systems analyst is writing a specification and the programmer is coding a program" is clearly of the {a AND b} form, though it means something different since there is presumably now no obvious connection between the specification and the program code.

2.3.4 The OR connective

Using similar notation to that used in Section 2.3.3 we describe the OR connective rule in the following way:

> The expression {a OR b OR c ... OR n} is FALSE only if all of a, b, c ... n are each of them FALSE.

Alternatively, we could put it this way:

> The expression {a AND b AND c ... AND n} will be TRUE if any one or more of a, b, c ... n is TRUE.

If you are beginning to sense a strong pattern running through our rule descriptions (we shall not call them definitions yet) then hang on to it. This is the first flavour of the *duality* characteristic of two-valued systems. Because of it, every valid logical picture of a rule can be turned into its 'negative print' by making a few simple switches of the wording.

Again a loud warning must be sounded about not confusing OR with the 'or' of everyday language and speech. The divide between them is perhaps even greater than in the AND case. We use 'or' to express strict alternative, sometimes signalled by a leading 'either'.

> You may use (either) COBOL or C (but not both).

This may be spoken without actually saying the explanatory words in brackets.

We also use 'or' for future conditional consequences (the time factor again), as in:

> (Either) You give me a cost-of-living increase or I resign.

It is also used sometimes with deliberate ambiguity such as in:

> Are you sick or tired (or both, or what)?

The OR of logic is intended, as the rule description above should indicate, to serve as meaning {a OR b OR both}. It is more fully named the Inclusive-OR, since it includes the case OR both (OR all, if more than two variables). This may seem uncomfortable if we are thinking about logic as an intelligent language (in the human sense), but logic must be clear and unambiguous to be properly formal, and also the OR as described is more useful in practice because it happens to be in dual relationship with AND.

2.3.5 The NOT connective

We use NOT to make explicit the reversal or switching between the two values of our two-valued system. It does not really connect in the sense of connecting two things together, but rather in the sense of attaching to a single variable in order to perform the switching function. We can describe it as follows:

> If a is TRUE, then (NOT a) is FALSE and
> if a is FALSE, then (NOT a) is TRUE.

To see it in simple use consider the expression:

> {a AND (NOT b)}

This expression is TRUE only if a is TRUE and b is FALSE! For example,

> The system failed and was not my design.

is TRUE only if the system did fail and at the same time it is FALSE to say that it was my design. You may well detect here a sneakily implied time connection (or even more). We will be cold logicians and ignore the possibility of hidden intelligence on this occasion, but you are welcome to start pondering about whether such subtleties should or should not be part of, say, the considerations in using an Expert System.

2.4 An introduction to logical inference

2.4.1 The importance of IF...THEN...

Appealing implicitly to common sense we have already made use of the IF...THEN... language syntax for handling inference as part of the ordinary processes of explanation. To programmers and systems analysts it is familiar also as part of programming language syntax. Its meaning may seem so obvious as not to require much explanation, and yet there is considerable potential confusion over its precise meaning in particular contexts, even among logicians themselves.

What seems obvious enough is the interpretation that, given the truth of

IF a THEN b

it follows that a being TRUE carries the implication that b also must be TRUE. But in everyday use of language we again find variety and even ambiguity in the use of this language construct. Consider the following:

If that's what you want, that's what you'll get; you're the user.
(Looking perhaps at a prototype screen layout design.)

Let us allow that the statement is indeed true, that is to say that the speaker is making a realisable promise and intends to honour it.

Let "this is what the user wants" be represented by w and "this is what the user will get" be represented by g. It is surely the case that w is TRUE implies that g is TRUE. But what if w was FALSE? First of all there seem to be two possible meanings to this. It could mean that what is being looked at is either something that the user has in fact rejected ("It's not what I want") or it is a screen layout not yet pronounced upon by the user ("No, I haven't said I want that", so the conclusion is in doubt).

In the real situation we may need to know quite a lot more about this little scenario before being sure which is the true meaning. At worse, we might find that the protagonists themselves are not sure or are in disagreement about it. The systems analyst may after implementation be heard to say that the screen layout is as it is because the user did not veto it.

What in more formal terms is in doubt here is whether the meaning should be

If w THEN g, AND IF NOT w THEN NOT g (2.1)

or

If w THEN g, AND IF NOT w THEN (g OR NOT g) (2.2)

The bracketed phrase in the second form (2.2) means something like: "a proposed screen layout not seen by the user may or may not appear in the final design."

Formal logic distinguishes between forms (2.1) and (2.2) as follows. It recognises (2.1) as being a case of value-equivalence between *w* and *g*. They are either both TRUE at the same time or both FALSE at the same time. Form (2.2) is the truly 'pure' inference, which ties the truth value of *g* to *w* only when *w* is TRUE. The form (2.1) is expressed more explicitly by mathematicians and logicians as **if and only if** (or **iff** for short).

Programmers should now recognise that the IF...THEN... syntax in procedural programming languages is of the equivalence or iff form. However, designers of logic-based programming languages would claim that they provide the pure inference syntax, but this is not always as clear as it may seem, as we shall discuss in Chapter 7.

2.4.2 The IMPLIES connective

Apart from the difficulties discussed above, the IF... THEN... format is a little clumsy for formal logic purposes, where a connective to represent it is preferred. There are a variety of symbols in use as we shall discuss in Section 2.5, but we shall use the word IMPLIES here. Thus the connective is placed between the variables in the same way as with AND and OR:

a IMPLIES *b*

Because this is not the iff interpretation it does not follow that this can be re-written as

b IMPLIES *a*

If it is not already clear why this is so, it should become clear when the actual logical operation of IMPLIES is looked at more precisely in Chapter 3.

This use of IMPLIES to represent inference is an essential brick in the building of knowledge processors. But so too is the use of the equivalence form, and the confusion between the two lies as a trap for the unwary, though usually within any one programming language or knowledge processing shell the confusion is avoided because of the constraints of the syntax. However, to put it at its mildest, it can do no harm to be conscious of the distinction. We shall pick up this discussion again in Chapter 3.

2.5 The algebraic representation of logic

2.5.1 The algebra and the calculus

The words 'algebra' and 'calculus' are often held in some awe by non-mathematicians. Let us try to defuse some of that feeling. We have already been using algebraic forms in earlier sections. 'Algebra' may be understood as another way of saying 'mathematical language'. Common algebra is a mathematical language for representing arithmetical and related systems.

Boolean algebra is one mathematical language for representing and communicating two-valued or binary systems.

As representations tend to be repetitious it becomes desirable to abbreviate, as we have done in Sections 2.3 and 2.4, by labelling propositions, switches, etc. by the use of a single character rather than with a word or phrase. We shall soon also be looking at single-character abbreviations for connectives. In this way algebraic representation soon starts to take on a classically mathematical appearance.

As we start to use the algebra to represent not just the basic elements of the system but also the reasoning processes that the system is amenable to, procedures and rules begin to emerge that themselves can be called-up and cited so as to save time and effort in writing down work. From this a calculus grows. The calculus will probably become the main working tool associated with the algebra and will be made tangible perhaps as a programming language or software package, though because of commercial or research considerations software packages often turn out to be 'dialects' of a calculus rather than straight representations of the pure textbook calculus. We shall in due course meet the Propositional Calculus for handling the propositional-statement type of logic and the Relational Calculus for handling the class-membership type of logic.

2.5.2 Alphabets and vocabularies

While having already introduced some of the vocabulary of Boolean logic (eg. AND, OR, NOT, IMPLIES, variable, value, expression) we have as yet only used the A to Z alphabet to depict the words. Even for everyday arithmetic this proves inadequate, and + for plus and − for minus, etc. are introduced to facilitate abbreviation. There is reasonably common currency in these symbols, though alternatives do exist and can lead to confusion in some contexts, eg. divide and multiply have tended towards / and * in computer-based arithmetic, though ÷ and × are still common in school arithmetic texts.

Although we have referred to Boole as perhaps the most important ancestor to modern logic (the Ancient Greek philosophers made significant initial progress too), the current logic scene is the arrival-point of many different research-work paths. Philosophers, mathematicians (both of the pure and applied variety), computer scientists, electronic engineers and linguisticians, followed in recent years by cognitive scientists and knowledge processing scientists, have along the way picked up different alphabetical habits and, worse still, sometimes different vocabularies. We have already referred to this Babelian situation in the opening chapter and now the time has come to face up to it.

We shall attempt to limit the complexity of this unfortunate problem by not giving a fully comprehensive account of the variety, but it is only right to give enough to act as a warning . Also to act as a possible encouragement, since there is more than a hint of suspicion that would-be readers of books and journal papers have abandoned the reading task under the misapprehension that they could not possibly understand the mathematics, when the problem has been 'merely' one of foreign alphabet. 'Merely' is in quotes, because looking at known words in strange alphabets is in itself no trivial problem as travellers to countries with foreign alphabets will know, where signposts may take a long time to read even when the place names required are known.

2.5.3 Some general terms

When we have considered, for example, the following:

> *a* AND *b*

we have referred to *a* and *b* as *variables*, AND as a *connective* and the whole as an *expression*. We might have prefixed each of these words with logical, Boolean or, sometimes, binary in order to clarify which algebra we are in, but we shall not do this where it is already clear enough from the context.

Alternatives for this vocabulary are given in Table 2.1. Often certain alternatives will be preferable because of the working context. If a thesis concerns propositional logic then it is more comfortable to use 'proposition' rather than 'variable', and 'compound proposition' rather than 'expression'. If wishing to emphasise that a variable is itself the simplest form of an expression, then calling it an 'atomic formula' might make sense.

The reader should study the table well if intending to browse through different textbooks. Each text will haves its own, probably very plausible, reasons for using the vocabulary it does but one can be almost guaranteed to experience some change in vocabulary when moving from any one text to another. We may in due time find a standard for the knowledge sciences becoming established. Here, we shall, by and large, use the terms at the *heads* of the columns in Table 2.1, and only resort to some of the alternatives where the context makes it very desirable to do so.

Table 2.1 *Alternative vocabularies in logic languages*

Variable	*Connective*	*Expression*
Proposition	Operator	Compound proposition
Simple proposition	Unary connective	Complex proposition
Elementary expression	(eg. NOT)	Statement
Atomic formula	Binary connective	Sentence
	(eg. AND)	Formula

N.B. (1) These are separate lists. There is not necessarily a horizontal grouping of preferred usages.

(2) The prefixes 'unary' and 'binary' clarify that the connective operates on one (eg. NOT *a*) and on two (eg. *a* AND *b*) variables respectively.

We shall want to use abbreviated forms for the truth-values TRUE and FALSE in order to save space and reduce apparent complexity. Table 2.2 shows the main alternatives available in contemporary texts. For general purposes we shall adopt the binary flags {1,0} (also called *binary digits* or more succinctly *bits*) for their relatively neutral appearance, but {T,F} and {Y,N} do have value in appropriate contexts.

It is also useful to add the word 'constant' to our vocabulary to use sometimes in preference to 'truth-value'. In other words, the pairs {1,0}, {T,F} and so on are all potential constants if they appear within expressions. The variables are so named because we want to indicate that they are capable of taking either of the truth-values (just as x may represent in common algebra a variable capable of taking any numeric value). When we decide that a variable must be constrained to be 1 or T or TRUE in a given circumstance, then it must remain constant to that value. It is therefore a kind of shorthand then to refer to the truth-values themselves as constants.

We shall also need to be able to symbolically represent the fact that one thing 'equals' or 'is the same as' another. We shall use the equals sign $=$ even when more rigorous mathematicians might resort to the three-barred 'identical to' symbol $\equiv$ to emphasise that one side of the sign is merely an equivalent form of the other, as in the trivial case $a \equiv a$, for example.

Round brackets (and) will be used in the usual way to hold together a sub-expression within a larger expression, as in

$$a \text{ AND } (a \text{ OR } b)$$

where it is understood that applying the OR connective will take precedence over the AND in the expression.

Where there is an intention to hold together the whole expression, either for readability or, more formally, to indicate that the whole needs to be regarded as a set of some kind, then the curly braces { } will be used.

Table 2.2 *Alternative symbols for logic values*

TRUE	FALSE
T (for True)	F (for False)
	$\perp$ (for inverted True)
Y (for Yes)	N (for No)
1 (the binary up-flag)	0 (the binary down-flag)

N.B. These can be read as pairings across the table.

We now turn to an even more confusing variety in relation to connectives. More confusing for two reasons. First, because special-character symbols look unfamiliar and 'weird' to the non-mathematically oriented, and, second, because the alternatives are quite extensive in number.

Table 2.3 shows the alternative symbol sets (and the alternatives within sets) which are in fairly common usage. However, other symbols might be encountered, especially as part of some proprietary software interfaces, and there is also a significant chance that the reader will come across sets made up of mixes other than from the sets shown in the table. Again, writers put forward the defence that in the context of certain specialised applications some sets are more 'standard' than others, but in fact there is no absolute and agreed standard in any particular field. So a writer or software designer's defence really is just that: "there is no standard, so I might as well do what I like and use what is most convenient" (a very important consideration in software design where the keyboard conventions for a family of software interfaces have become familiar to the user).

Our selection for this book is based on the following arguments:

& for AND — because the name of this symbol, ampersand, is a good reminder of the function name.

∨ for OR — because it is the most commonly encountered. The V-shape of the symbol is called vel, the Latin for 'or'!

Table 2.3 *Alternative symbols for logic connectives*

AND	OR	NOT	IMPLIES
AND (a, b)	OR (a, b)	NOT (a)	Footnote (i)
$a \wedge b$ or $a \& b$ or $a.b$	$a \vee b$	$\neg a$ or $\tilde{\ } a$	$a \rightarrow b$ or $b \leftarrow a$
$a.b$	$a + b$	$\bar{a}$ or a'	Footnote (ii)
$a \cap b$	$a \cup b$	$\neg a$ or $\tilde{\ } a$ or a'	$a \supset b$ Footnote (iii)

N.B. (i) This set is associated with user-friendly software. The IF (a, b) would mean 'equivalence' (if and only if) in such contexts and not 'implies'.

(ii) This set is often found in engineering texts, where 'implies' is usually not dealt with.

(iii) This horseshoe symbol for IMPLIES is unfortunate in that it faces the opposite way to the subset symbol (see 5.2.3).

~ for NOT	because, although ⌐ is equally common, the tilde is more commonly found on wordprocessing keyboards. The prime, or single quote placed immediately after the variable (eg. *a'*), is admittedly more common in this latter respect, but it can be confused with its other uses in mathematics. Prime also follows the variable, which does not read as well as a sign which, like NOT itself, precedes that which it operates upon.
→ for IMPLIES	because the arrow is a graphic representation of direction of effect, a very important feature of implication and inference. The reversal of the direction of the arrow can be used consistently to reverse the inference where needed. The double-headed arrow ↔ can be used for equivalence again with some consistency of meaning. We avoid the 'horseshoe' symbol at all costs (see Table 2.3, footnote (iii)).

2.6 Tabular and graphic representations of logic

2.6.1 Introduction

It may seem undiplomatic at this juncture to introduce yet more variety, but the variety to be reviewed here is of a much more cheering kind. It provides not only additional tools to facilitate working in logic but also helps with the true understanding of 'what is going on' in logic processes. For the systems analyst it also provides some interesting potential analysis and design tools for someone who is already well used to using charting tools as part of the repertoire. As with most instances of using pictorial material it should also lighten the reading and learning load for those who find a page of written or algebraic logic tough on the powers of concentration. There are a lot of us in that boat!

2.6.2 The truth table

An important characteristic of two-valued systems is that, for situations where there are only a few variables, it is feasible to contemplate setting out all possible evaluations of an expression. This is not the case in common algebra, for instance in Table 2.4, where the table of values is of infinite length unless bounds are set to the range of values and the type of values to be tabulated, eg. for positive integer (whole-number) values of x and y between 0 and 9 inclusive, which even so would show a hundred rows!

Table 2.4 *Tabulation for the sum of two numbers*

x	y	$x + y$
0	0	0
1	0	1
2	1	3
...	...	...

Since Boolean variables can only ever take one of two values, Boolean tables—conventionally called **truth tables**—have finite numbers of rows. With two variables there can only be four rows, with three variables eight rows..., with n variables $2 \times 2 \times ... \times 2$ (n times) rows. Although an initial reaction might be 'So what?', the setting out explicitly of truth tables can in fact prove useful and instructive. For instance, the complete truth tables for two-variable expressions of the common connectives AND, OR and NOT are shown in Tables 2.5, 2.6 and 2.7 respectively. We have here used our adopted conventions for the symbol alphabet. It is a useful exercise, if such a presentation is unfamiliar, to work through the rows of these tables satisfying yourself that they are in accordance with our descriptions of the connectives given in Section 2.3.

Table 2.5 *Tabulation of the AND connective*

a	b	$a \& b$
0	0	0
0	1	0
1	0	0
1	1	1

Table 2.6 *Tabulation of the OR connective*

a	b	$a \vee b$
0	0	0
0	1	1
1	0	1
1	1	1

Table 2.7 *Tabulation of the NOT connective*

a	$\tilde{a}$
0	1
1	0

2.6.3 The K-map

For two-variable cases we can set out the table as a coordinate two-dimensional map of the evaluations as shown in Fig 2.6 in two alternative layouts for {a & b}. This format is called a **Karnaugh map**, or K-map for short, named after its deviser. Although based on a straightforward enough idea it yields, as we shall see in Chapter 4, some remarkable power to a logic modeller. Layout (ii) helps in the ease with which that power can be tapped, but is slightly more difficult to read since it is not always clear how to interpret the edge readings, using as they do the NOT connective as part of the label. We shall deal with this in Chapter 4.

The reader should now draw the K-map (in either format) for the case {a ∨ b}. The only difference from Fig 2.6 should be in the pattern of the four binary digits below and to the right of the lines.

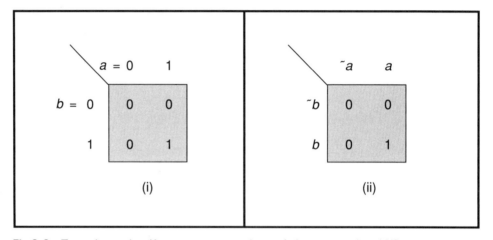

Fig 2.6 Two alternative K-map representations of the connective AND

2.6.4 The decision tree

If a diagram is constructed based on the idea of testing the truth of variables and branching accordingly, we arrive at the familiar tree diagram as shown in Fig 2.7. Viewed as a logic structure (Fig 2.7(i)) it has a strangely 'inverted' form and, although tree diagrams are excellent illustrators of what its user should do in certain circumstances (as illustrated by Fig 2.7(ii)), it is a useful logic modelling tool only in certain special circumstances which will be discussed in later chapters. In Fig 2.7 we have reverted to logic 'words' for the sake of readability.

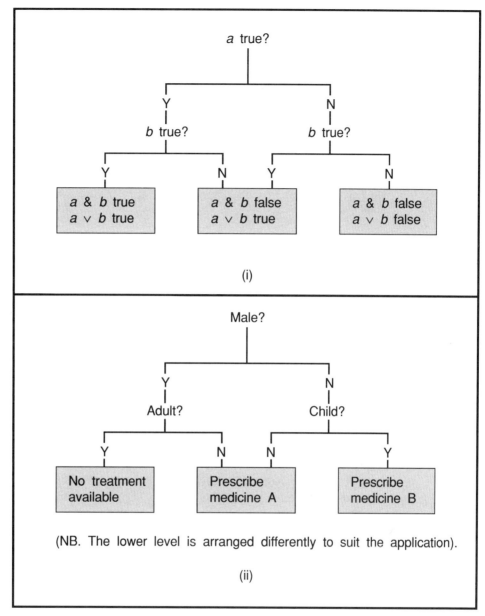

Fig 2.7 (i) A decision tree for two variables; (ii) Example of an actual tree

2.6.5 The decision table

The decision table format for a system of interconnected decisions is a special one geared to aiding the analysis of such a system, especially in respect of its self-consistency and its apparent incompleteness. Even in the current scene of rapidly evolving software tools for the aid of problem analysis it is still a relatively important paper tool for the systems analyst.

Decisions	Rules			
Is patient male?	Y	Y	N	N
Is patient adult?	Y	N	Y	N
Prescribe A		X	X	
Prescribe B				X
No treatment available	X			

Fig 2.8 The decision table equivalent of Fig 2.7(ii) (Note the assumption that NOT adult = child.)

We shall deal with it at some length in Chapter 4 and then in further detail in Chapter 6.

Figure 2.8 illustrates the format for the same decision set as appeared in the decision tree in Fig 2.7(ii). For both the decision tree and the decision table it is unlikely that the user is explicitly interested in the Boolean expressions describing the actions to be taken. Thus the expressions at the bottom of the inverted tree in Fig 2.7(i) are not typical of what one normally expects to see in an actual application. The expressions at the bottom of the tree in Fig 2.7(ii) are much more likely to be useful. In Fig 2.8, the logical expressions have been omitted to emphasise this point; the action labels A, B, C, and D would be descriptions, or refer to descriptions, of the actions to be taken like those in Fig 7.2(ii).

However, it is becoming increasingly useful to understand the mapping between formal logic and both these formats in analysis work which leads towards knowledge processing systems. This question is explored further as the book develops.

2.6.6 The Venn diagram

The conventions for Venn diagrams are far from being well-defined and formalised. They were devised as aids to visualising the logic of class membership, especially in the branch of mathematics known as set theory (ie. the theory of sets). The setting up of a Venn diagram might be described as follows:

1 Draw a rectangular frame to contain the Universal Set, ie. the set containing all things which are under consideration.

2 For each class under consideration draw a circle to contain all things which belong to that class. Ensure that each circle overlaps every other circle to provide sub-areas containing things which belong to more than one class.

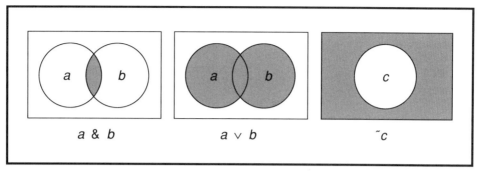

Fig 2.9 Shading used for illustrating sets

3 Indicate by shading, marking or inscription the sub-set that is required to be visualised.

Following this procedure for two sets a and b, Fig 2.9 shows, by shading, the sub-areas for $\{a \& b\}$, $\{a \vee b\}$ and $\{\tilde{~}c\}$ respectively.

It follows from statement 1 above that the universal frame is in one sense a representation of the constant 1 or T, since the question as to whether any 'thing' in the system is enclosed within it is inevitably always going to be YES. The constant 0 or F, on the other hand, is a vanishingly small circle since it encloses not a single one of the 'things'.

2.6.7 Black-box charts

An alternative way of viewing a logical expression is to regard it as a network of input–output processes. Here each connective is a process; the inputs are variables in their pre-process state and the outputs the variables in their post-process state. How the process of change is actually achieved is 'hidden inside an opaque box', hence the use of the term 'black-box'. The representation of the operation of a single connective such as $\{a \& b\}$ as shown in Fig 2.10(i) is therefore just one box and might seem trivial. But for more complex expressions the diagram or chart shows the progressive build-up of how the final result is achieved. (See Fig 2.10(ii).)

This type of chart is very well suited to the needs of the computer circuit design engineer whose concern is chiefly with what goes on between input and output in a logic circuit, that is to say a circuit whose function is to transform binary digits in one state into binary digits in another state. The physical manifestations of these binary digits are tiny electronic pulses which are either above ($=1$) or below ($=0$) a measurable threshold value. One such transformation is illustrated in Fig 2.10(ii) by the inclusion of a sample input in brackets accompanying each annotation.

A word of warning, though. Engineers have devised a visual language for such charts, giving each connective a specially shaped box so as to avoid having to label each box separately.

We shall not use this diagrammatic language at any juncture but

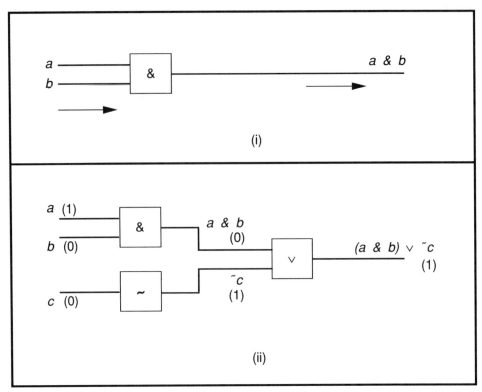

(i)

(ii)

Fig 2.10   Logic charts using black-boxes

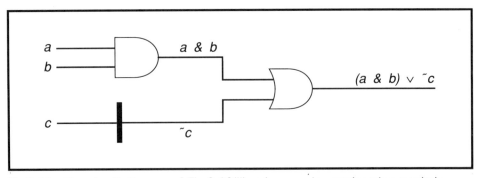

Fig 2.11   The circuit diagram of Fig 2.10(ii) redrawn using engineering symbols

Fig 2.10(ii) is reproduced in Fig 2.11 to give an idea of the general appearance of this specialist type of chart. Engineers also use the jargon 'gate' when referring to the black-box units. Thus there are AND-gates and OR-gates, etc.

2.6.8   Indentational formats

Although tables and diagrams provide a useful alternative perspective from which to view the structures of logic, they are not the most commonly experienced way of writing down logic processes. Writing (and especially program coding) is a sequential procedure of stating first one thing, then another, then another, and so on. As the use of interactive screens becomes increasingly the mode of working for programmers and systems analysts alike, especially with fourth generation tools, there is admittedly a certain loosening of this constraint. But nevertheless the end product is still almost inevitably a sequential list of statements about the logic. It is useful therefore to have a representational form which recognises this fact while still aiming to provide some of the advantages of diagrammatic formats.

A common solution is to present the logic as a list of statements top-to-bottom of page, reading each line left-to-right (ie. the conventional Western reading mode), but to indicate the structure of the logic by varying the indentation from the left of each successive paragraph. This is usually familiar enough to programmers and systems analysts as structured English

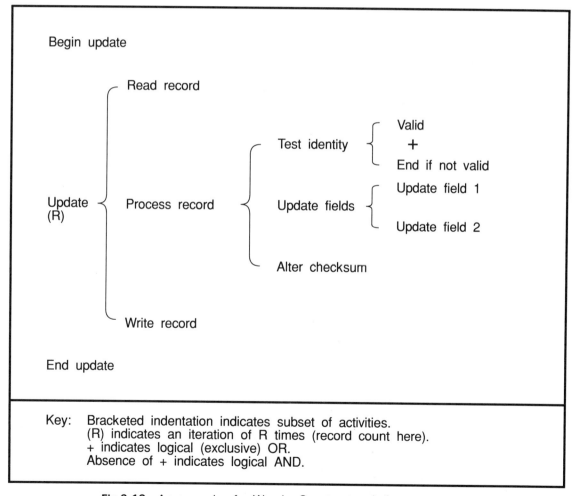

Fig 2.12 An example of a Warnier-Orr structured diagram

or pseudocode, or sometimes in a form even closer to diagrammatic appearance such as in the Action Diagramming from the James Martin stable or the structured diagrams of the Warnier-Orr methodology. See Fig 2.12.

2.6.9 More formal tree and network diagrams

Although it should be clear that the representations we have looked at in Sections 2.6.2 through 2.6.7 do have some degree of usage convention about them, they all fall short of being truly formal languages. Formal representations have been devised for the representation of formulae and of knowledge processing structures. These may end up as tree-like or as generalised network-like diagrams depending upon the purpose to which they are being put. We shall look at some of these in Chapter 9 when we shall be in a better position to understand the principles at issue.

2.6.10 Other representations

The systems analyst makes use of a wide range of tabular and charting conventions, some of which may be clearly recognisable in the representations we have been discussing. Decision trees and tables may be used in both the analysis of existing decision processes and in rationalising the re-design of these for program specification purposes or system documentation. They might even be used as instructional models in users' manuals or in work-room wall charts. Venn diagrams, without necessarily being given that label, may well be used in system descriptions to clarify how classes of objects may share or not share those objects. Truth tables, K-maps and black-box logic charts are less likely to have been encountered in this context, except perhaps as part of courses of study for professional and other examinations. A better understanding of the common underlying logic of all these may well serve to enhance the charting repertoire of the systems analyst however.

The question arises as to whether flowcharts, data structure diagrams, data flow diagrams, and so on, have any connection with logic. Although they are tools designed expressly to help the systems analyst capture the logic of analysis and design work, the connection is not necessarily a direct one but we shall need to leave the question until a stage when we can tie them up to logical mapping in general, in Chapter 4 and elsewhere.

It is not our intention to turn every one of the systems analyst's charting tools into a version of logic representation. But it is perhaps worth putting our case this way: if anything can show what is the root connection between all these tools it is worth knowing about it. Whether two-valued logic is totally adequate to this purpose is very doubtful, but, as we have already stated, it is a usefully simple start.

2.7 Moving between representations: an example

2.7.1 Introduction

It should be clear by now that the different representations have degrees of usefulness related to the context of use, so there is inevitably some inappropriateness about representing the same example in all of the different ways for an unspecified context. However, in order to provide something 'to bite on' we shall carry out this artificial exercise by way of a demonstration. Since it relates most easily to an everyday situation we shall use the propositional statement context as our base example.

The example propositions are as follows:

a The computer is switched on.
b The printer is switched on.

The single-connective compound propositions might therefore be expressed:

$\{a \& b\}$ Both the computer and the printer are switched on. (2.3)

$\{a \lor b\}$ One or both of the computer and printer are switched on. (2.4)

$\{\tilde{\ }a\}$ The computer is not switched on. (2.5)

There are other ways of putting these compound propositions into natural language statements, but we should try to avoid obvious ambiguity such as "Either the computer or the printer is switched on" in place of (2.4), because it is unclear whether or not it means both are switched on.

2.7.2 Interpreting the propositional forms

If both a and b are TRUE (abbreviation T for this will now be used), then statements (2.3) and (2.4) are also both T, while (2.5) is FALSE (F). In fact (2.5) is easily dealt with, for if a is F, ie. it is not true to say that the computer is switched on, then (2.5) is clearly T.

If a and b are both F, then statements (2.3) and (2.4) are also both F, which again is not difficult to accept.

We have to be careful however that we agree with the situation that obtains when one or other of the two equipments is on while the second is off. Because we were careful with the wording of (2.3) and (2.4) it should be relatively easy to accept that in this instance (2.3) is F while (2.4) is T. Less-precise formulations of (2.4) such as mentioned in Section 2.7.1 above do not interpret so clearly.

Let us look at a two-connective expression $\{a \lor \tilde{\ }b\}$ and consider what this means. It is already getting difficult to avoid clumsiness in natural language even though the logic is absolutely unambiguous:

$\{a \lor \tilde{\ }b\}$ Either the computer is switched on or the printer is switched off or both these circumstances obtain. (2.6)

49

It might be easier to understand this if re-phrased as

> {˜(˜a & b)} It is not the case that the computer is switched off and
> at the same time the printer is on. (2.7)

But are (2.6) and (2.7) really the 'same'? Yes, they are! You can get there
by common-sense interpretation, and we shall see in the next chapter how
we can get there by manipulation of the algebra also. But first let us look
at some of the other representational forms to see how they fare with our
simple example.

2.7.3 The Venn diagram interpretation

We can interpret the Venn diagram circles as representing enclosures of
points which stand for different situations. Thus all occasions when a is on
will be represented by a point within the circle labelled a and similarly for
b. Occasions when both are on will thus fall within the overlap of the two
circles and occasions when they are both off will fall outside the two circles
but within the Universal frame. The frame here encloses all possible
occasions involving the switch-states of a and b.

By progressively shading-in the areas covered by the states a, $˜b$ and
finally $\{a \vee ˜b\}$ we arrive at Fig 2.13(iii). Similarly by doing this for $\{˜a\}$,
$\{b\}$, $\{˜a \& b\}$ and finally $\{˜(˜a \& b)\}$ we arrive at Fig 2.14(iv). From these
two representations the sameness is visually obvious.

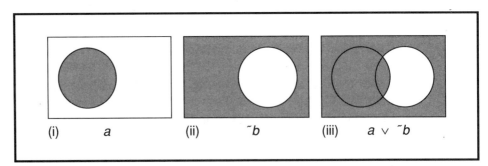

| (i) | a | (ii) | $˜b$ | (iii) | $a \vee ˜b$ |

Fig 2.13 Identifying $a \vee ˜b$ on a Venn diagram

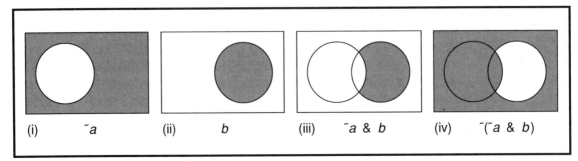

| (i) | $˜a$ | (ii) | b | (iii) | $˜a \& b$ | (iv) | $˜(˜a \& b)$ |

Fig 2.14 Identifying $˜(˜a \& b)$ on a Venn diagram

2.7.4 Truth table and K-map interpretations

The truth tables (Fig 2.15) and the K-maps (Fig 2.16) make the same point as the Venn diagrams by exhibiting identical bit-patterns for each of the two forms (2.6) and (2.7). With the truth table the progressive evaluations to the final column help to show how the results are derived and the working is therefore checkable. The K-maps simply show the identical results and are not really very useful for a demonstration example of this kind.

a	b	˜b	a ∨ ˜b	˜a	˜a & b	˜(˜a & b)
1	1	0	1	0	0	1
1	0	1	1	0	0	1
0	1	0	0	1	1	0
0	0	1	1	1	0	1

Fig 2.15 Truth table evaluation for a ∨ ˜b and for ˜(˜a & b)

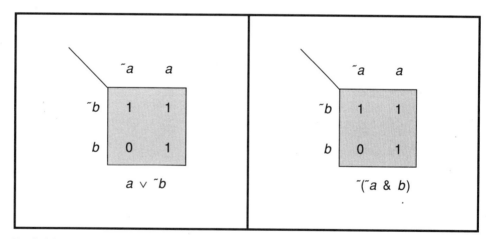

Fig 2.16 Identical K-maps for a ∨ ˜b and for ˜(˜a & b)

2.7.5 Decision tree and table interpretations

The tree format is perhaps rather unexpectedly difficult to use to illustrate the identical nature of the two forms (2.6) and (2.7). Figure 2.17 attempts it but is awkward to read and decipher. The decision table, although ostensibly modelling the forms from the same point of view (decisions), is much clearer however. A little patient study will show that Fig 2.18 does present a fairly straightforward demonstration of the identity.

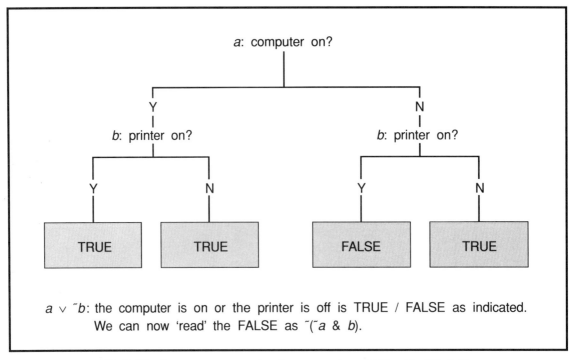

$a \lor \tilde{}b$: the computer is on or the printer is off is TRUE / FALSE as indicated. We can now 'read' the FALSE as $\tilde{}(\tilde{}a \& b)$.

Fig 2.17   A tree representation for comparing $a \lor \tilde{}b$ with $\tilde{}(\tilde{}a \& b)$

a: Is computer switched on?	Y	Y	N	N
b: Is printer switched on?	Y	N	Y	N
$a \lor \tilde{}b$: Computer on or printer off	X	X		X	←
$\tilde{}a \& b$: Computer off and printer on			X	
$\tilde{}(\tilde{}a \& b)$: Negation of above line	X	X		X	←

Fig 2.18   Decision table comparison of $a \lor \tilde{}b$ with $\tilde{}(\tilde{}a \& b)$

2.7.6   Black-box interpretation

We see from Fig 2.19 that this representation, rather like the decision tree, allows us to see the two paths leading to the same result. The difference, which may or may not be an advantage depending on context, is that it shows only the required path in each case. This enables us to focus on the function under consideration but does not show both paths on the same diagram nor how the path fits into the overall pattern of all possible paths.

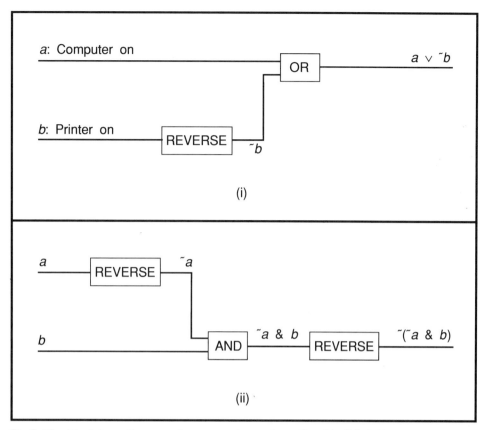

Fig 2.19 Black-box diagrams for comparison of $a \lor \tilde{\ }b$ with $\tilde{\ }(\tilde{\ }a \& b)$

2.7.7 Switch circuit interpretation

Although it might have seemed best to start with this since our propositions were about equipment being switched on or off, this is not at all the case. What has to be coped with here is that we are at different language levels even though they are both switch circuit languages. The $\{\tilde{\ }b\}$ means that when the printer is switched on it is in fact off in its connection with the computer; in other words it is in the opposite state to its labelled function b! Either the connection with the computer is a different circuit from the power supply or we are not talking about a circuit at all perhaps. The statement $\{a \lor \tilde{\ }b\}$ might be purely to do with rules of procedure—what is permitted, rather than what is physically constrained. This is an important initial lesson in the dangers of poor knowledge representation and we shall avoid bad practice by not showing the example as a switch circuit at all. Trying to illustrate the printer's perverse role could test beyond reason our illustrative powers!

3 Some Basic Rules of Logic

3.1 Elementary manipulation procedures

3.1.1 Logic as modelling

The study of representational alternatives in Chapter 2 should have driven home the point that logical expressions, whether algebraically described or shown as tables or diagrams, are models of some actual or imagined reality. We build models so that we can experiment with reality without necessarily changing it. How meaningful the experimentation is will depend heavily upon the quality of various types of mapping as follows:

- The mapping from reality to our conceptualisation of it.
- The mapping from our conceptual intentions to the logic model itself.
- The mapping from the logic model to its practical implementation, which can range from a pen-and-paper model to a sophisticated computer-driven knowledge base.
- Finally, there is the question of the quality of the experimentation itself and the interpretation of it. (It is not difficult to imagine a good-quality knowledge base being badly used to produce some very poor-quality proposals for action.)

In this chapter we are going to look at the rules of logic which, if observed, will mean that our experimentation with the logic model will be sound and should also assist us in some of the design issues implied by the first three of the above.

3.1.2 Basic rules

A rigorous mathematical treatment of logic needs painstakingly to trace the establishment of the formal language to ensure that a truly complete and self-consistent system is emerging. We shall avoid this level of meticulousness here in the interest of ease of understanding, but nevertheless point the reader to the fact of the matter because, if for no other reason, such treatment may be encountered in most specialised texts which may be used as follow-up reading to this one. The building-up of a system of reasoning in mathematical terms is fascinating in that amazingly powerful and complex methods of working evolve out of what at the earliest stages appear to have almost nothing to offer.

The plethora of terms by which rules of different kinds are known include such terms as theorem, property, law, principle, equivalence, as well as rule itself. All these in one way or another refer to a proven transformation which the reasoner using the system can quote and thereby achieve a short-cut path to some desired goal of proof or demonstration. A correct and valid use of the rules is itself always a potential rule since it is then a proven transformation, ripe for quotation in future reasoning. Programmers and systems analysts are familiar enough with the idea in the form of tested sub-programs which can be incorporated into future programming. However, the difference in that arena is that 'tested' falls a good deal short of the rigour of 'proven', in spite of much recent and current research into achieving program provability.

Hofstadter deals entertainingly with the question of how one ever gets started in this painfully rigorous process in his truly magisterial and yet readable book by referring to the rule from which all others derive as the Fantasy Rule. The general reader might explain it as 'stating the obvious' or as some equivalent common-sense description, but the point is that even stating the obvious has to be checked to see that it is a reliable rule that can be adopted as the foundation of all that follows!

We shall start a little way down the line from the Fantasy Rule and present as our initial and basic rules ones that are nevertheless, we hope, 'obvious'. If they are to be quoted in further reasoning they must be labelled in some way and it is therefore useful, though not essential, to adopt the classical names as used by the logicians. These names can appear as adjectives or as nouns, eg. the commutative law or commutation, depending upon context.

Whether we call our rules just that, that is to say **rules**, or **laws** or whatever, will not matter too much at this stage. In general, a correct setting-out of reasoning to **prove** a law can be termed a **theorem** and a collection of rules relating to a particular theme might be called **properties**, but interchanging these terms will not be considered a serious crime in this text.

Two laws of commutation

$$a \& b = b \& a$$
$$a \vee b = b \vee a$$

Two laws of association

$$a \& (b \& c) = (a \& b) \& c$$
$$a \vee (b \vee c) = (a \vee b) \vee c$$

Three laws of negation

$$a \,\&\, \tilde{a} = 0$$
$$a \vee \tilde{a} = 1$$
$$\tilde{(}\tilde{a}) = a$$

Two laws of identity

$$a \,\&\, a = a$$
$$a \vee a = a$$

(These are known formally as the idempotent laws, but *laws of identity* sounds more straightforward.)

Four properties involving the constant values 1 and 0 are those which we

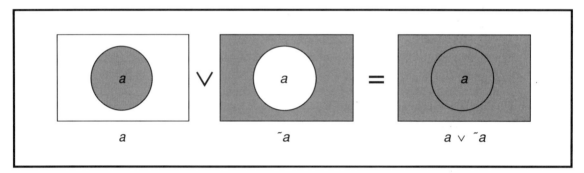

Fig 3.1   Venn diagram illustration of $a \vee \tilde{a}$

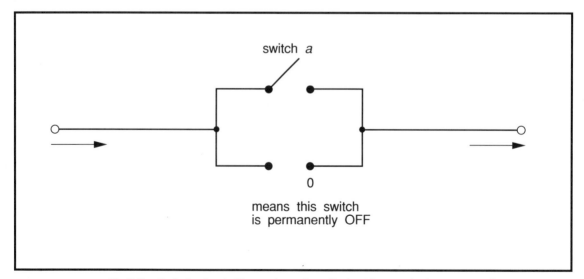

Fig 3.2   Switch circuit illustration of $a \vee 0$ showing equivalence to a alone

shall refer to simply as the **four constants properties** in future:

$$a \,\&\, 1 = a$$
$$a \,\&\, 0 = 0$$
$$a \lor 1 = 1$$
$$a \lor 0 = a$$

If you find any of these less than obvious you should take the opportunity of testing your understanding using a variety of the representation methods presented in the previous chapter. For example, Fig 3.1 shows $\{a \lor \tilde{\ }a = 1\}$ illustrated by Venn diagram and Fig 3.2 shows $\{a \lor 0 = a\}$ illustrated by a switch circuit.

3.1.3 The feasibility of evaluation

In Section 2.6.2 it was pointed out that two-valued logics have finite numbers of evaluations for any given expression. Looked at from the point of view of the truth table this means that for a two-variable case there are four (two to the power of two) rows of values in a fully evaluated table, for the three-variable case eight rows (two to the power of three), for the four-variable case sixteen (two to the power of four), and so on. For these lower numbers of variables it is therefore possible to contemplate looking at the full evaluations to see what happens to an expression under all possible circumstances of the value-combinations of its variables. As the number of variables gets greater this is no longer the case; the information explosion soon overtakes us and we find we need the help of computers to cope. However, even the *inference engine* software of a knowledge processing system such as a large expert system may find that being called upon to carry out evaluations involving literally hundreds of variables is simply beyond its realistic capabilities, since two to the power of hundreds takes it into the realm of astronomic number counts such as the number of atoms in the Galaxy!

However, there is considerable merit in learning about these evaluations at the human scale of things since this can give us insight into what is going on at the computer and even galactic scales. We shall attempt this mostly at the two- and three-variable scale, with an occasional foray into the four-variable scale.

3.1.4 Evaluation precedences

When we formulate compound expressions we have to be careful that its interpretation as a formula is unambiguous. For example, when we write

$$\{a \,\&\, b \lor c\}$$

do we intend that $a \,\&\, b$ is to be evaluated first and the result then used to evaluate the $\lor c$, or should the $\lor$ connective be given priority? In other words, using brackets to clarify the distinction, which of the following two interpretations is intended:

$$\{(a\,\&\,b)\vee c\} \quad \text{or} \quad \{a\,\&\,(b\vee c)\}?$$

Does it matter? Are they not perhaps the same anyway? The answer in this case is that they are not. The two distributive laws which we shall meet in Section 3.1.5 demonstrate this, but we can appeal to common-sense to illustrate that they are not. Consider, for example:

> You must work the Friday and Saturday shifts together or the Sunday shift on its own (or all three if you wish).

The bracketed clause is required because this is the logical inclusive OR. This is definitely not the same as

> You must work the Friday shift together with either the Saturday or the Sunday shifts (or all three if you wish).

The constraint in the first interpretation is that Friday and Saturday shifts must be paired, while in the second it is that the Friday is an imperative shift.

It should be clear from this that the brackets are necessary to avoid the confusion. Some authors allow the & connective precedence over the ∨ by convention; in other words the first interpretation would apply if brackets were omitted. We shall use brackets in any case.

However, brackets are not necessary if the connectives are the same throughout the expression, because of the laws of commutation. The expressions

$$\{a\,\&\,b\,\&\,c\} \quad \text{and} \quad \{a\vee b\vee c\}$$

are each unambiguous since the same result comes from evaluating either of two connectives first. For example,

> You must take Friday, Saturday and Sunday off. (The & case)

> You can take off Friday, Saturday or Sunday, or any combination of the one to three days. (The ∨ case)

are not affected by giving any one combination of days precedence over another.

Brackets are necessary, though, if the NOT connective is to be applied to the result of other connective operations. Thus,

$$\{(\tilde{\,}a)\,\&\,b\} \quad \text{is not the same as} \quad \{\tilde{\,}(a\,\&\,b)\}$$

For example,

> Exclude the appendix, i.e. $(\tilde{\,}a)$, and include the bibliography, i.e. $\&\,b$.

is not the same as

Exclude both the appendix and the bibliography, i.e. $\tilde{\ }(a\,\&\,b)$

However, because the NOT connective applies to only one variable and does not actually connect two variables, we can permit the convention to drop the brackets in the first case above. That is to say, we write

$\tilde{\ }a\,\&\,b$ for $(\tilde{\ }a)\,\&\,b$

without loss of clarity. Even $\tilde{\ }\tilde{\ }a$ is permissible (one of the sides of the equality in the law of negation), though it might sometimes be better to write $\tilde{\ }(\tilde{\ }a)$ in some circumstances for clarity's sake. Though the brackets are redundant they are not incorrectly present and one is well advised, when in doubt, to include brackets to clarify or emphasise the intended interpretation.

We can now formalise the precedence procedure as follows:

- Evaluate within brackets first.
- Within brackets, or where there are no brackets, evaluate the NOT connective ($\tilde{\ }$) first.
- Evaluate sub-expressions containing either all & or all ∨ connectives last.

How to apply this procedure in practice is shown in the next section.

Before proceeding to that section it is worth pausing to wonder what happens to the equivalent of these precedence rules and the use of brackets in natural language communications. The answer is that often misunderstandings do occur. Absolute clarity of interpretation, such as in legal writing, requires a rather unnaturally styled prose, with many clauses and emphases. It is not unusual, even then, to hear of individuals exploiting loopholes in the law to interpret to their own advantage and in a way not intended by the original legal rules writer. In face-to-face communication we find the need to resort to such devices as repetition, modulations of voice and gesture, and emotive signalling in order to clarify even our most logical statements.

This is just another little boost to the message that the representation of knowledge is by no means a straighforwardly simple and logical task. The expert will tell us that this is partly because we have in this text as yet only looked at the crudest of tools, namely two-valued logic. This is a valid point, but note well the 'partly'!

3.1.5 Full evaluation of expressions

We have already demonstrated the full evaluation of the single-connective expressions in Section 2.6.2 when introducing truth table representation. We now do so for more complex expressions and will use the method as a way of proving two further laws which are basic but not at all obvious. These are

Two laws of distribution

$$a \& (b \lor c) = (a \& b) \lor (a \& c) \tag{3.1}$$

$$a \lor (b \& c) = (a \lor b) \& (a \lor c) \tag{3.2}$$

These are called 'distributive' in recognition of how the initial connective operations $a\&$ and $a\lor$ spread through the terms within the brackets. (This might bring to mind the distributive law of multiplication in common algebra: $a(b + c) = ab + ac$. But the difference with logic is that it applies to both AND and OR connectives, while in common algebra it is true for multiplication alone.)

If we fully evaluate each of the four sides of these two laws we shall expect the bit-value pattern to be identical for each of the two parts. First, a further point needs to be made about the setting-up of evaluation tables. Since these expressions involve three variables we can anticipate there being eight rows. It is useful in helping us later to make useful pattern recognitions of the symbols to set out these rows in a systematic way. The best system in this respect is to start with two blocks of four of each value for the variable in the leftmost column (see Table 3.1(i)). Then, for each block, set out two blocks of two for each value for the next variable (see Table 3.1.(ii)). This leaves alternating values to be inserted in the third column (see Table 3.2).

The result prevents errors arising when trying to count out the eight sets of row values by trial and error and provides a useful convention that ensures every truth table is alike. (Note that the convention could vary in that the 0 blocks could be placed above the 1 blocks initially and in that the blocking could start in the rightmost rather than in the leftmost column.)

Table 3.1 *Entering the set-up values in a truth table*

a	b	c	...	a	b	c	...
1				1	1		
1				1	1		
1				1	0		
1				1	0		
0				0	1		
0				0	1		
0				0	0		
0				0	0		
	(i)				(ii)		

Table 3.2 *Evaluation of b ∨ c*

a	b	c	b ∨ c	a & (b ∨ c)		b ∨ c
1	1	1				1
1	1	0				1
1	0	1				1
1	0	0	0	0		0
0	1	1				1
0	1	0				1
0	0	1				1
0	0	0	0	0		0
			(i)			(ii)

The next stage is to add evaluation columns so that the steps of evaluation are carried out in the correct sequence according to the precedences discussed earlier. Thus, for the left-hand side of law (3.1) which we will evaluate first, the within-brackets must take precedence, yielding the layout shown in Table 3.2(i). In the Table we have also carried out the first step of evaluating the column for $b \vee c$. Rather than evaluating each row one at a time, we note that this is an OR connective and use the definition that the outcome of an OR operation is 0 only if both of the variables is 0. Having worked with the economy of 'spotting the exceptional case' we can then go ahead and fill in the remaining values as ones (see Table 3.2(ii)).

The completion of the final full-evaluation column should be tackled in a similar manner. Since this is an AND operation the first three rows should be entered as 1 because the columns for a and $b \vee c$ are both 1 in only these three cases. The remaining entries are then 0, and the completed table of evaluation is as shown in Table 3.3.

Following a similar procedure we can produce the full evaluation table for the right-hand side of (3.1), as shown in Table 3.4.

Table 3.3 *Completed evaluation of a & (b ∨ c)*

a	b	c	b ∨ c	a & (b ∨ c)
1	1	1	1	1
1	1	0	1	1
1	0	1	1	1
1	0	0	0	0
0	1	1	1	0
0	1	0	1	0
0	0	1	1	0
0	0	0	0	0

Table 3.4 *Evaluation of $(a\,\&\,b) \vee (a\,\&\,c)$*

a	b	c	$a\,\&\,b$	$a\,\&\,c$	$(a\,\&\,b) \vee (a\,\&\,c)$
1	1	1	1	1	1
1	1	0	1	0	1
1	0	1	0	1	1
1	0	0	0	0	0
0	1	1	0	0	0
0	1	0	0	0	0
0	0	1	0	0	0
0	0	0	0	0	0

The bit patterns in the final columns of both Table 3.3 and 3.4 are indeed identical and we can therefore deduce that the two sides of (3.1) are also identical. This will seem a somewhat crude and long-winded method of proof to the mathematically inclined but there is something satisfying about being able to literally see the truth of the identity in such explcit detail.

We now carry out the same exercise for law (3.2), but set out the two evaluations in the same table. The colums tagged LHS (for left-hand side) and RHS in Table 3.5 are identical, again proving the law.

A more compact method of setting out these evaluations is to insert the bit values under the connective in the expression to be evaluated as in Table 3.6, which is Table 3.5 re-formatted in this way. It is not so easy to read through the working since it does not follow a left-to-right column sequence.

Computer science has, of course, long recognised this general problem of the 'reading' and 'computational' sequences not matching (it happens in common algebra and arithmetic too) by devising so-called **reverse notations**. Such notations recognise the sequence in which symbols need to be picked up so as to progress through the computation in one direction smoothly.

Table 3.5 *Evaluation showing the truth of law (3.2)*

				LHS			RHS
a	b	c	$b\,\&\,c$	$a \vee (b\,\&\,c)$	$a \vee b$	$a \vee c$	$(a \vee b)\,\&\,(a \vee c)$
1	1	1	1	1	1	1	1
1	1	0	0	1	1	1	1
1	0	1	0	1	1	1	1
1	0	0	0	1	1	1	1
0	1	1	1	1	1	1	1
0	1	0	0	0	1	0	0
0	0	1	0	0	0	1	0
0	0	0	0	0	0	0	0

Table 3.6 *Evaluation showing the truth of law (3.2)*

a	b	c	$a \vee (b \& c)$		$(a \vee b)$	&	$(a \vee c)$
1	1	1	1	1	1	1	1
1	1	0	1	0	1	1	1
1	0	1	1	0	1	1	1
1	0	0	1	0	1	1	1
0	1	1	1	1	1	1	1
0	1	0	0	0	1	0	0
0	0	1	0	0	0	0	1
0	0	0	0	0	0	0	0

        ↑LHS   ↑RHS

For example if we adopted the notation:

connective(variable1, variable2,...)

we could represent the RHS of (3.2), appearing in the rightmost column heading in Table 3.6, as

$$\& (\vee (a, b), \vee (a, c))$$

It is called 'reverse' because the picking-up takes place from right to left. Although this has the advantage of clarity for computing machines, it does seem less comfortable to the human mind-and-eye combination, which by and large prefers to look at wholes while interpreting parts.

We conclude this section with a more complex example of evaluation to help clarify and extend understanding about the value of carrying out a full evaluation when this is feasible.

EXAMPLE   In the following description by a salesperson of current rules of practice concerning a bookstand display we have inserted logical symbols in a cumulative way (within the square brackets) to express what we believe (we could be wrong) is the correct model of what is being said.

> We never [~] display our Young Readers' books [y] or books on Zoology [z] when in Xanadu [x] for political reasons. [Thus so far we have $\{ {\sim}((y \vee z) \& x))\}$.] Mind you, we will always display non-zoological Young Readers' books [$y \& {\sim}z$] wherever the exhibition. Thus making finally:

$$\{(y \& {\sim}z) \vee {\sim}((y \vee z) \& x))\}$$

The sequence of evaluation needs to be as shown in the column heads to Table 3.7, which is also a full evaluation of the expression derived above. We have used A, B, etc. to represent sub-expressions in order to save space in the column heads. (Such substitution is in any case very sound algebraic practice.)

Table 3.7 *Evaluation of the bookstand logic*

x y z	$y \vee z$ $(=A)$	$A \& x$ $(=B)$	$\tilde{}B$	$\tilde{}z$	$y \& \tilde{}z$ $(=C)$	$C \vee \tilde{}B$ $(=$ full expression$)$
1 1 1	1	1	0	0	0	0
1 1 0	1	1	0	1	1	1
1 0 1	1	1	0	0	0	0
1 0 0	0	0	1	1	0	1
0 1 1	1	0	1	0	0	1
0 1 0	1	0	1	1	1	1
0 0 1	1	0	1	0	0	1
0 0 0	0	0	1	1	0	1

What is the significance here of the bit values in the final fully-evaluated column? Since the salesperson was speaking of permissions to display, we can take 1 to mean 'can display with the conditions of x, y, z as shown' and 0 to mean 'can't display'.

It would seem sufficient to say that display of Zoology books in Xanadu is not allowed. So what happened to the salesperson's constriction on Young Readers' books? If we have interpreted correctly, this was overridden by the second sentence. The logical evaluation is certainly correct. But if fed back to the speaker we may well hear "No, that's not what I meant at all", even though that is indeed what was said. Back to the drawing board. Well, at least back to the salesperson to go through the rules more carefully.

But the example does illustrate in a semi-serious way not only that people do not communicate with the language of formal logic but also that the investigating systems analyst might be able to use logic to clarify both the user's and their own understanding of 'complex' procedures and rules.

Again we must remind ourselves that the tools we have used here are fairly crude ones, but when we look at more sophisticated ones later we may then be in a better position to use them intelligently because of an understanding which has grown out of the simpler foundations.

3.1.6 Deriving models from tables

In the example above we tried to formulate the logical model as an expression emerging out of a textual statement, as it were. Another approach, and one more familiar to engineers, is to slot values into a truth table (or some equivalent representation) and then derive the algebraic model from that. An engineer, for example, may know what an electrical circuit is intended to do under various conditions but needs the complete model on which to base the actual design.

We shall demonstrate this by using a probably unrealistically simple example but one with sufficient design in it to make the point.

EXAMPLE   An alarm circuit is required which can be controlled from the use of three switches. One switch is to be a master switch, switching the circuit on and off whatever the state of the other two. Otherwise the circuit should be switched on if any two (or all three) of the switches are on.

We label the three switches a, b and c, with c being assigned the role of the master overriding switch. Rather than trying to interpret the above as an algebraic expression direct, we draw up the truth table as shown in Table 3.8(i). Ones and zeros here represent the switches and the circuit being in ON and OFF states respectively. The bits in the Alarm circuit column have been inserted by carefully interpreting the requirements described above. Thus whenever c is 1, the circuit is 1 also. Whenever two out of three of a, b, or c are 1, the circuit is 1 also. (Some cases will already have been covered by the first statement of course. As long as there are no contradictions we are alright.)

How can we now derive the correct expression for the whole function of this circuit? One method is to write down, for each row where the function is 1, the algebraic description, called a **minterm**, of truth states of the variables in that row. Thus the first row might be read as saying that, for the circuit to be on, we could switch a AND b AND c on. OR we could achieve it by having a AND b on AND c off (second row). The algebraic forms of these have been entered in Table 3.8(ii).

Each of these minterms is an alternative way of ensuring that the circuit is on, therefore the full algebraic expression needs to be the connection of them by using OR connectives throughout:

$$(a \& b \& c) \vee (a \& b \& \tilde{c}) \vee (a \& \tilde{b} \& c) \vee (\tilde{a} \& b \& c) \vee (\tilde{a} \& \tilde{b} \& c) \qquad (3.3)$$

This is a correct model in the sense that the logic does work. However it is

Table 3.8   *Modelling the alarm circuit*

a	b	c	Alarm circuit	Minterms
1	1	1	1	$a \& b \& c$
1	1	0	1	$a \& b \& \tilde{c}$
1	0	1	1	$a \& \tilde{b} \& c$
1	0	0	0	—
0	1	1	1	$\tilde{a} \& b \& c$
0	1	0	0	—
0	0	1	1	$\tilde{a} \& \tilde{b} \& c$
0	0	0	0	—
		(i)		(ii)

not a correct design in that there is a far simpler form which can be arrived at via various means of simplification. The circuit built from the above formula would be rich in redundant circuit connections! We urgently need to look to simplification then to help us, and this we shall do in the next section.

3.2 Algebraic simplification of expressions

3.2.1 Introduction

In the rules presented in Section 3.1 we have some indication as to what simplification might be like in practice. In a sense, simplicity of form is relative to what one needs the formula for, but it seems probable that $\{0\}$ may be regarded as simpler in form than $\{a \& \tilde{\ }a\}$ for example (one of the rules of negation), that $\{a\}$ is simpler than $\{a \vee a\}$, and so on. The laws of commutation, association and distribution, however, seem to offer ways of showing the same expression in different, but equally simple, forms. Use of these as transformations, however, can often lead to simplification opportunities that were previously hidden. For example,

$$
\begin{aligned}
& a \& (\tilde{\ }a \vee b) \\
= {} & (a \& \tilde{\ }a) \vee (a \& b) \quad \text{(Distribution)} \\
= {} & 1 \vee (a \& b) \quad \text{(Identity)} \\
= {} & a \& b \quad \text{(Constants)}
\end{aligned}
$$

The words in brackets act as justification for each step just completed by appealing to the rule or law which we agree can be used without proof because we have established it in some previous working.

3.2.2 Further simplification-aiding rules

Let us look at some further rules (laws, properties) which offer other opportunities of simplification or transformation.

Four laws of adsorption

$$
\begin{aligned}
a \vee (a \& b) &= a \\
a \& (a \vee b) &= a \\
\\
a \& (\tilde{\ }a \vee b) &= a \& b \\
a \vee (\tilde{\ }a \& b) &= a \vee b
\end{aligned}
$$

Two De Morgan rules

$$
\begin{aligned}
\tilde{\ }(a \& b) &= \tilde{\ }a \vee \tilde{\ }b \\
\tilde{\ }(a \vee b) &= \tilde{\ }a \& \tilde{\ }b
\end{aligned}
$$

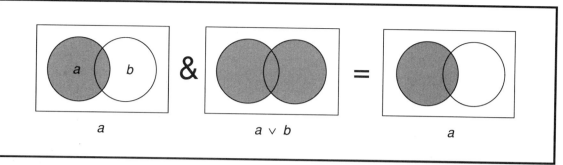

Fig 3.3 Venn diagram illustration of an adsorption law

Table 3.9 *Demonstrating one of the De Morgan rules*

a	b	a & b	~(a & b)		~a	~b	~a ∨ ~b
1	1	1	0		0	0	0
1	0	0	1		0	1	1
0	1	0	1		1	0	1
0	0	0	1		1	1	1
		(i)				(ii)	

The adsorption laws are potentially very useful in simplification, the right-hand sides being obviously simpler in each case than the left-hand sides. The name 'adsorption' should be taken as meaningful and aiding in memorising the laws if we notice that in each of them the first term 'absorbs' all (first two laws) or part (second two) of the term within brackets.

De Morgan's rules are not so obviously helpful, but in fact the RHSs are more useful for reasons we shall come to later.

Again we shall bypass formal proofs (though the perspicacious reader may well have spotted a proof of one of the adsorption laws just a few paragraphs before we stated them!), but we can always resort to the use of our other representional forms to help if we wish to satisfy ourselves that they are indeed true. For instance, Fig 3.3 shows how the Venn diagram might be used to this effect for one of the adsorption laws, and Table 3.9 shows the truth table being used for this purpose for one of the De Morgan rules.

EXAMPLE   A computer-system access-control protocol is defined as being: "If the user's code interprets as Grade I manager [call this Boolean variable g] access, permission is to be granted. Permission is to be granted for non-Grade I managers [$~g$] so long as they are not non-Grade I managers with special-restriction codes [call this last category s].

Let us interpret this as being expressible as

$$g \vee (\tilde{g} \,\&\, \tilde{\;}(\tilde{g} \,\&\, s))$$

Note that this means we take both 'so long as' and 'with' here to mean 'at the same time', ie. AND.

Simplification proceeds as follows:

$$
\begin{aligned}
& g \vee (\tilde{}g \,\&\, \tilde{}(\tilde{}g \,\&\, s)) \\
={}& g \vee (\tilde{}g \,\&\, (\tilde{}\tilde{}g \vee \tilde{}s)) && \text{(De Morgan)} \\
={}& g \vee (\tilde{}g \,\&\, (g \vee \tilde{}s)) && \text{(negation)} \\
={}& g \vee (\tilde{}g \,\&\, \tilde{}s) && \text{(distribution)} \\
={}& g \vee \tilde{}s && \text{(distribution)}
\end{aligned}
$$

We could surmise from this that the control could be more simply expressed as: "Grant access to Grade I managers. Also grant access to all others without special restriction."

It might seem that we have used logic here merely to clean up some tortuous English, which should have been possible using native common-sense and a good ear for clear English. That's as may be in this example. More complex and even more tortuous examples might prove beyond our everyday use of language interpretation. Of course, if the English is tortuous in the extreme, there will also be difficulty in arriving at a 'correct' logical formulation in the first place, but the attempted analysis may help in highlighting the obscurities, and certainly the simplification result should provide fruitful material for lively debate with the formulators of the original rules!

3.2.3 The principle of duality

Almost without exception we have presented the rules (laws, properties) in pairs or quadruples. In fact any rule can be transformed into its *dual* (a kind of twin) by the interchange of

AND   with OR and vice-versa
1   with 0 and vice-versa

The only exception is the rule $\{\tilde{}\tilde{}a = a\}$ which contains none of the four features involved in duality.

The reader is left to inspect the laws introduced so far in order to confirm the principle of duality.

The principle applies to any result so long as both sides of an equation are submitted to the transformation. Take the example at the end of the previous section. The simplification could be expressed as the equation:

$$ g \vee (\tilde{}g \,\&\, \tilde{}(\tilde{}g \,\&\, s)) = g \vee \tilde{}s \tag{3.3} $$

There is a dual result to this one which is

$$ g \,\&\, (\tilde{}g \vee \tilde{}(\tilde{}g \vee s)) = g \,\&\, \tilde{}s \tag{3.4} $$

Duality runs through two-valued systems in other ways also and can yield some surprising short-cuts in algebraic working under certain circum-

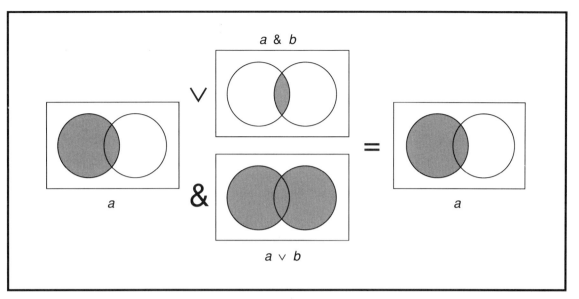

Fig 3.4 Venn diagram illustration that $a \lor (a \& b)$ is the same as $a \& (a \lor b)$ and that both are simply equal to a

stances. However, the newcomer to these ideas should beware the trap of overenthusiasm in applying the principle. What the principle enables one to do is to generate from one equating of forms, say $A = B$, a new equating of forms $C = D$. It does *not* aver that either $A = C$ or $B = D$. In (3.3) and (3.4) above this is more clearly seen by looking at the two RHSs which are fairly obviously not identical. However if one side of an equation remains unchanged under a duality transformation then identity of form *is* implicit. For example, the two simpler versions of the laws of adsorption:

$$a \lor (a \& b) = a$$
$$a \& (a \lor b) = a$$

are duals each of the other, the RHS remaining unchanged by the transformation. Since for this $B = D$ is true, it must also be true that $A = C$, ie.

$$a \lor (a \& b) = a \& (a \lor b)$$

which is indeed a true identity of forms. The Venn diagram in Fig 3.4 shows that this is so. The dual of this newly discovered rule is itself, ie. the same rule, the effect of transforming it being merely a swapping over of sides of the equation. (But note the warning that we were looking here at a special case, partly in order to emphasise the fact that the equating of $A = C$ and $B = D$ is not the general case.)

3.2.4 Rules of factoring

In common (arithmetic) algebra, factorising:

$$ac + ad + bc + bd = (a + b)(c + d)$$

and the reverse of the process, multiplying out, play important roles in providing useful algebraic transformations. How does factorising work in Boolean algebra, if at all? The answer is that they are very similar because of the way the distributive laws work (which are actually special cases of 'factorising' transformations anyway). If we replace multiplication with & and summation with ∨ in the common algebra rule above we get a valid Boolean law, and duality immediately gives us a second one as well:

Two laws of factoring and de-factoring

$$(a \& c) \vee (a \& d) \vee (b \& c) \vee (b \& d) = (a \vee b) \& (c \vee d) \qquad (3.5)$$

$$(a \vee c) \& (a \vee d) \& (b \vee c) \& (b \vee d) = (a \& b) \vee (c \& d) \qquad (3.6)$$

The de-factoring, working from the RHS to the LHS, is a question of taking all possible pairings between the terms within each bracket. The switch circuit illustration of the first of these is shown in Fig 3.5.

These laws are taking us into moderately difficult algebra but special cases of them (eg. the adsorption laws) are not so difficult and two in particular

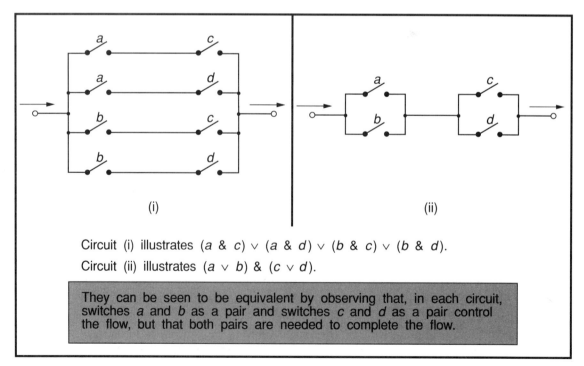

Circuit (i) illustrates $(a \& c) \vee (a \& d) \vee (b \& c) \vee (b \& d)$.

Circuit (ii) illustrates $(a \vee b) \& (c \vee d)$.

They can be seen to be equivalent by observing that, in each circuit, switches *a* and *b* as a pair and switches *c* and *d* as a pair control the flow, but that both pairs are needed to complete the flow.

Fig 3.5 Switch circuits used to illustrate the de-factoring rule (3.5)

are very powerful simplifying agents. Because it is the de-factoring direction that provides the simplification effect, we state these as de-factoring from left to right:

Two special rules of de-factoring

$$(a \& b) \vee (a \& \tilde{~}b) = a \tag{3.7}$$

$$(a \vee b) \& (a \vee \tilde{~}b) = a \tag{3.8}$$

This surprising reduction (bear in mind that the RHSs here are equivalent to the LHSs in the full versions of the laws given as (3.5) and (3.6)!) comes about because of the repetition of variables in the two factors and the subsequent effect of the laws of identity and constants in the multiplied-out expression.

It is tempting to use the thought that the b and $\tilde{~}b$ have 'cancelled each other out' in (3.7) and (3.8) as an *aide-memoire* for these two rules, but great care needs to be taken in avoiding the use of such cancelling when it is *not* applicable! For example, it is *not* permissible to cancel in the following case:

$$(a \& b \& c) \vee (a \& \tilde{~}b)$$

De-factoring this expression produces

$$a \vee (c \& \tilde{~}b)$$

The Venn diagrams for the de-factoring rules will make the truth of them very evident, but for variety let us look at the switch circuit models shown in Fig 3.6. Since the switches b and $\tilde{~}b$ must always be in opposite states

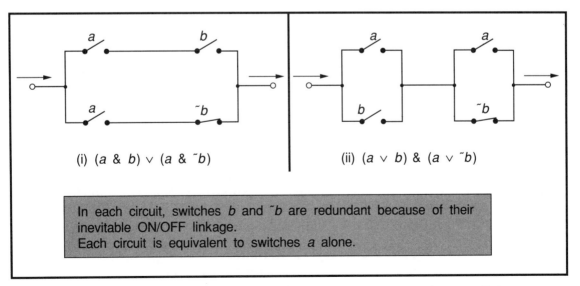

(i) $(a \& b) \vee (a \& \tilde{~}b)$ (ii) $(a \vee b) \& (a \vee \tilde{~}b)$

In each circuit, switches b and $\tilde{~}b$ are redundant because of their inevitable ON/OFF linkage.
Each circuit is equivalent to switches a alone.

Fig 3.6 Switch circuits used to illustrate the special rules (3.7) and (3.8) (think of a rigid link which forces this complementary state of affairs between the two switches) their combined effect is null and the whole circuit behaves as if it comprised the switch a alone.

3.2.5 Normal forms

Finally, in this glimpse of the basics of algebraic simplification, we consider another type of transformation which is not always as obvious a case of simplification as we have met so far. As the name implies, *normalisation* of forms in any subject is to do with presenting them in a standard and recognisable way which, although it may not always produce shorter or less-complex-looking forms, does simplify their handling in a variety of ways. (Normalisation of relational forms of data is another case in point.)

There are two normal forms that we are interested in at this stage, each based on one of the two connectives AND and OR. First we should mention that, although we have not used the terminology so far, the two connectives are sometimes referred to as the *conjunctive* and *disjunctive* operators respectively. These are awkward words, but their purpose is to draw attention to the 'confining' aspect of AND and the 'expansive' aspect of OR. (Look at the Venn diagrams for these connectives in Fig 2.9 to see what this means.) The two normal forms are, as a consequence, called the conjunctive and disjunctive normal forms respectively.

Conjunctive normal form is achieved by factoring an expression as a conjunctive string of sub-expressions which have no conjunctions within them. These sub-expressions are usually referred to as the *clauses* of the whole expression (ie. a special case of factors) for reasons that will become apparent later. Thus a generalised picture of this form, with c representing a clause, is

$$c_1 \& c_2 \& c_3 \& \ldots \& c_N$$

Remember that because of the commutation law we do not need to bracket any of these connective pairs, since sequence of evaluation is immaterial. However, as we saw in the previous section, brackets will be needed around some of the clauses to clarify their boundaries.

An example of a conjunctive normal form is

$$(a \vee \tilde{}b) \& \tilde{}c \& (a \vee c) \& b$$

This has four clauses, the second and fourth of which do not need brackets because they are single-variable sub-expressions. Note that though this is a normal form it is not in its simplest form. Simplification as follows can be used to reduce it to the simplest normal form:

$(a \vee \tilde{}b) \& \tilde{}c \& a \& b$	(Adsorption)
$a \& \tilde{}c \& a \& b$	(Adsorption)
$a \& \tilde{}c \& b$	(Identity)

Disjunctive normal form is, of course, the dual of conjunctive normal form. It comprises a disjunctive string of sub-expressions which contain no disjunctions within them. Consider the following example, given first in non-simplified form and then simplified as with the previous example:

$(a \& b) \lor \~b \lor (a \& c) \lor \~c$
$(a \& b) \lor \~b \lor a \lor \~c$ (Adsorption)
$a \lor \~b \lor a \lor \~c$ (Adsorption)
$a \lor \~b \lor \~c$ (Identity)

These two examples are deceptively similar. The truth is that they are both examples of expressions which have very simple simplified normal forms. Each comprises only a single sub-expression when viewed in its 'other' (dual) normal form. The disjunctive normal form of the first example is

$(a \& \~c \& b)$

and the conjunctive normal form of the second is

$(a \lor \~b \lor \~c)$

the brackets being used here only to emphasise that each is a single clause.

Let us consider a more complex example by returning to our model derived as a disjunction of minterms, Form (3.3) in Section 3.1.6. We can understand now that this is indeed in disjunctive normal form, each minterm being in fact a clause of the expression. Let us use algebraic simplification on this as follows:

$(a \& b \& c) \lor (a \& b \& \~c) \lor (a \& \~b \& c)$
    $\lor (\~a \& b \& c) \lor (\~a \& \~b \& c)$
$= ((a \& b) \& c) \lor ((a \& b) \& \~c) \lor (a \& (\~b \& c))$
    $\lor (\~a \& b \& c) \lor (\~a \& (\~b \& c))$ (New brackets)
$= (a \& b) \lor (\~b \& c) \lor (\~a \& b \& c)$ (De-factoring)
$= (a \& b) \lor (\~b \& c) \lor ((\~a \& b) \& c)$ (New brackets)
$= (a \& b) \lor ((\~b \lor (\~a \& b)) \& c)$ (Factoring)
$= (a \& b) \lor ((\~b \lor \~a) \& c)$ (Adsorption)
$= (a \& b) \lor (\~(a \& b) \& c)$ (De Morgan)
$= (a \& b) \lor c$ (Adsorption)

which is the simplest disjunctive normal form.

By factoring this form we can put it into its simplest conjunctive normal form:

$(a \lor c) \& (b \lor c)$

Remember that the original exercise was the design of a circuit and this we can now do with optimum efficiency. We select the disjunctive normal form because it happens to have no repeated switch reference and is therefore simpler to set out as a circuit. (See Fig 3.7.)

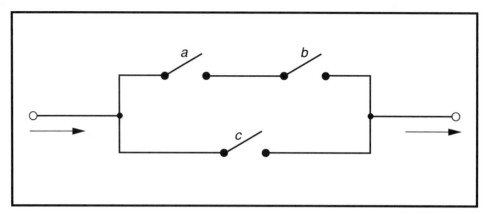

Fig 3.7 The final design for the required circuit (see example in text)

3.3 The algebra of inference

3.3.1 Equivalence

A glance back to Section 2.4 will remind us that there are (at least) two interpretations of what we mean by an IF...THEN... inference. One of these interpretations was 'if and only if' (or iff) which means that in the form:

IF (and only IF) *a* THEN *b*

the truth values of *a* and *b* must always be identical. We can show this in the truth table representation by making TRUE (=1) both rows where the variables are of the same value (see Table 3.10).

The table has been annotated with comments referring to typical interpretations of IF...THEN...ELSE in computer programming syntax. Notice the awkward change of tense for the middle two rows. This indicates that a temporal logic is in action; the first and last rows give the logic of what must happen because of the truth (=1) of the intended operation of iff, whereas the middle rows describe what obtains (ERROR) after the operation turned out not to have worked as it should (=0). We can illustrate the potential trap in talking about the logic of programming by considering

Table 3.10 *The iff truth table*

a *b*	iff *a* then *b*	*Programming meaning*
1 1	1	*a* is TRUE, so perform *b* (ie. *b* must be TRUE)
1 0	0	*a* is TRUE, *b* not performed, therefore ERROR
0 1	0	*a* is FALSE, *b* performed, therefore ERROR
0 0	1	*a* is FALSE, perform not *b* but something ELSE

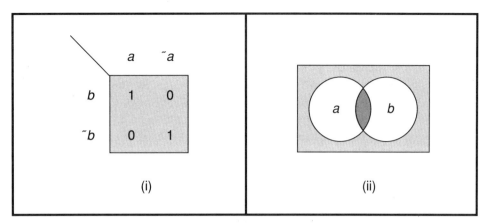

Fig 3.8 Equivalence shown as (i) a K-map and (ii) a Venn diagram

the two statements:

- The program produces a report of monthly sales.
- The program is designed to produce a report of monthly sales.

The first statement may prove to be false on certain occasions whereas the second proving false is a more permanently serious matter!

Since it is a complete truth table it effectively defines a connective relationship between the two variables. Again, a variety of symbols will be encountered in the literature, but we will use the most common of them—the triple-barred equals sign $\equiv$. Commutative and associative laws apply to this connective.

$$a \equiv b = b \equiv a$$
$$(a \equiv b) \equiv c = a \equiv (b \equiv c)$$

By identifying the minterms from the truth table we can formulate equivalence as a Boolean expression in terms of AND, OR and NOT. The minterms for the first and fourth rows are $\{a\,\&\,b\}$ and $\{\tilde{\ }a\,\&\,\tilde{\ }b\}$, thus the full expression must be

$$(a\,\&\,b) \vee (\tilde{\ }a\,\&\,\tilde{\ }b)$$

which will not simplify any further. This ability of the three common connectives to be always sufficient in modelling any other connectives we may care to devise is termed a **basis** by mathematicians. The set {AND, OR, NOT} form a basis for two-valued systems but they are not the only set that do so as we shall see in Section 3.4

As to other representations of equivalence, Fig 3.8 shows a K-map and a Venn diagram for the connective. Other representations do not tell us any more than these.

3.3.2 Implication

We have already introduced IMPLIES (or $\rightarrow$) as a connective and have made the point that if $\{a \rightarrow b\}$ is true it does not follow that $\{b \rightarrow a\}$ (which we can also write as $\{a \leftarrow b\}$, ie. a is implied by b) is necessarily true. The truth table for the operation of the connective for each case makes this clear (Table 3.11) since the bit-value pattern for each is different.

We see that this implication-type of inference is less constraining than equivalence since, when a (called the *antecedent* in $\{a \rightarrow b\}$) is FALSE, it does not constrain b (called the *consequent*) to be FALSE also, as equivalence would. It is important to grasp this point clearly. Failure to do so is the source of much confusion in interpretations of formal logic.

EXAMPLE   Here are similar statements, one of which is equivalence, the other implication:

A: If the password presented is acceptable, access will be granted (Equivalence)
B: If the program fails the test, it must contain an error.   (Implication)

Statement A represents equivalence because the intended meaning is (let us say) that access is grantable if and only if the password is acceptable at the point of its being tested by the access program. If password is OK, access granted; if not OK, access not granted.

But statement B is not saying, should the program not fail the test, that means it does not contain an error, as any experienced programmer will readily recognise. There may be any number of errors not detected by the test, which may, for example, have been designed for the purpose of detecting one particular type of error.

Of course we always have to make assumptions about *intended* meanings. Which is why we inserted the 'let us say' in parentheses in the explanation of statement A above. If the intended meaning was that the password was only one means of granting access, then it is no longer equivalence, since 'password not OK' may lead on to other tests which may after all grant access. Also statement B is clearly based on the assumption that the test is a sound one, that the program will not fail the test due to the test itself being in error!

Table 3.11 *Comparison of $a \rightarrow b$ with $b \rightarrow a$*

a	b	$a \rightarrow b$	$b \rightarrow a$
1	1	1	1
1	0	0	1
0	1	1	0
0	0	1	1

The basic Boolean expression for implication should help to clarify these points. If we gather the minterms from the truth for $\{a \to b\}$, we have

$$(a \& b) \vee (\tilde{~}a \& b) \vee (\tilde{~}a \& \tilde{~}b)$$

This will simplify as follows:

$$= b \vee (\tilde{~}a \& \tilde{~}b) \qquad \text{(De-factoring)}$$
$$= b \vee \tilde{~}a \qquad \text{(Adsorption)}$$
$$= \tilde{~}a \vee b \qquad\qquad\qquad\qquad\qquad (3.5)$$

The natural-language wording for this is "either a is false or b is true, or both", which is not a particularly clear message.

However, if we consider the case for the falseness rather than the truth of the implication, we would have just the single minterm:

$$a \& \tilde{~}b \qquad\qquad\qquad\qquad\qquad (3.6)$$

which reads "it is not possible for b to be false while a is true", which somehow is easier to understand. Note, by the way, that the expression for the falseness (3.6) is the De Morgan transformation of the truth (3.5) of the implication:

$$\tilde{~}(\tilde{~}a \vee b)$$
$$= \tilde{~}\tilde{~}a \& \tilde{~}b$$
$$= a \& \tilde{~}b$$

which is as it should be since we have, in effect, simply used the negation rule:

$$\{\tilde{~}(\tilde{~}x) = x\}$$

where x here is the statement as to whether the implication is true or not.

The Venn diagram for the implication expression (see Fig 3.9(i)) yields another way of describing implication, namely "if it's true for a it has to be true for b too", and it is sometimes helpful to illustrate this by drawing the special Venn diagram shown in Fig 3.9(ii). We shall use this illustrative form in later chapters.

If for a pair of variables (a, b) it turns out that implication is true in both directions, ie. $\{a \to b\}$, and at the same time $\{b \to a\}$, we find that this brings us back to equivalence. The algebraic proof of this is

$$(a \to b) \& (b \to a)$$
$$= (\tilde{~}a \vee b) \& (\tilde{~}b \vee a) \qquad\qquad \text{(using 3.5)}$$
$$= (\tilde{~}a \& \tilde{~}b) \vee (\tilde{~}a \& a) \vee (b \& \tilde{~}b) \vee (b \& a) \qquad \text{(de-factoring)}$$
$$= (\tilde{~}a \& \tilde{~}b) \vee 0 \vee 0 \vee (b \& a) \qquad\qquad \text{(constants)}$$
$$= (a \& b) \vee (\tilde{~}a \& \tilde{~}b)$$

which is the expression for equivalence, as Table 3.12 shows.

But it also feels right that two-way implication should be the same as

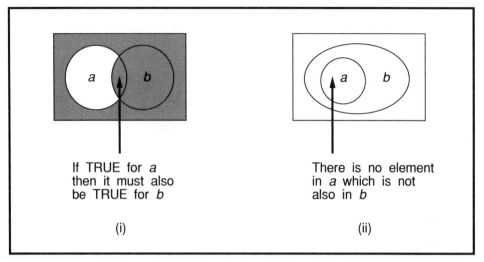

Fig 3.9 Venn diagram illustrations of implication $a \rightarrow b$: (i) using standard diagram and (ii) using special diagram

equivalence, bearing in mind what we have said about the distinction between the meanings of equivalence and implication.

Finally, we should mention that an example of implication in a temporal mode is that of *causation* or cause-and-effect. In causation, the event a guarantees that the event b will follow, but the event b could occur as a result of some other event also. We can see this quite literally in program running, where an event—say the display of an error message that a disk is full—may be the result of one of any number of different causing-events.

But temporal logic throws up some important difficulties and differences of meaning. Once the trigger event has actually happened in real time, the reverse implication becomes determined and what was only an implication before the triggering becomes an equivalence after it. (Sherlock Holmes understood this distinction very well, using causation to trace back from evidence while exploring the possible implications which would provide the traceable paths.) This again throws our thoughts forward to issues of artificial knowledge processing, for the triggering of deductive events is

Table 3.12 *Deriving the algebraic expression for equivalence*

a	b	$a \leftrightarrow b$	Minterms
1	1	1	$a\,\&\,b$
1	0	0	
0	1	0	
0	0	1	$\tilde{\ }a\,\&\,\tilde{\ }b$

Therefore $a \leftrightarrow b$ is expressible as $\{(a\,\&\,b) \vee (\tilde{\ }a\,\&\,\tilde{\ }b)\}$

exactly like this. (The jargon has it that 'a rule fires'.) Knowledge process software lies sleeping with many possible logical implications within bosom, but an activated version of it gets committed to real effects. Whether or not the software will be capable of emulating Sherlock Holmes and retracing its deductive path in the reverse direction will depend on, amongst other things, the sophistication of the software. These are issues to return to later.

3.4 Other connectives

We have been able to define each binary (two-variable) connective perfectly satisfactorily by means of the four-row truth table. Each of the four connectives AND, OR, IMPLIES and EQUIVALENCE has shown a unique bit-pattern in the result column, and we have a fifth pattern if we include IS IMPLIED BY, ie. $\{b \leftarrow a\}$. But there are altogether sixteen different possible bit patterns for four bits; are there therefore eleven further possible connectives? Yes, there are!

We shall not be looking at all of these, though logic texts do often list and name them. There are two, however, which are of special interest in the field of computer technology and these are called NAND and NOR. We have already made mention in Section 3.3.1 of the concept of a basis in a mathematical system, this being a set of operational functions which are sufficient to express all operations, including any of those other than the base set. Thus we have seen how EQUIVALENCE and the two IMPLIES connectives can be expressed in terms of AND, OR and NOT. The basis concept is of value to computer logic circuit designers since they know that, given a stock of these types of gate (as they call an electronic logic connective—see 2.6.7), they can design and build potentially any functional circuit.

But the remarkable property of both NAND and NOR is that *they each form a basis on their own*. This means that the circuit designer can contemplate building NAND-only or NOR-only circuits. Cheaply produced boards containing hundreds of such gate-arrays need only appropriate connecting-up to create custom-designed circuits.

The truth table definitions of these two versatile connectives are shown in Table 3.13.

Pattern-spotting should reveal that they are the negative images of AND and OR respectively, and hence their names. It should be noted that operationally speaking NAND is AND followed by NOT, and NOR is OR followed by NOT, even though the N for NOT has been pre-fixed to the names. However, the algebraic notation we have adopted, ie. the preceding tilde (~), makes the readability consistent. The representations of NAND and NOR in terms of our previous basis connectives are therefore:

$$a \text{ NAND } b = \tilde{\ }(a \& b) \qquad a \text{ NOR } b = \tilde{\ }(a \lor b)$$

Table 3.13 *The NAND and NOR truth tables*

a	b	a NAND b
1	1	0
1	0	1
0	1	1
0	0	1

(i)

a	b	a NOR b
1	1	0
1	0	0
0	1	0
0	0	1

(ii)

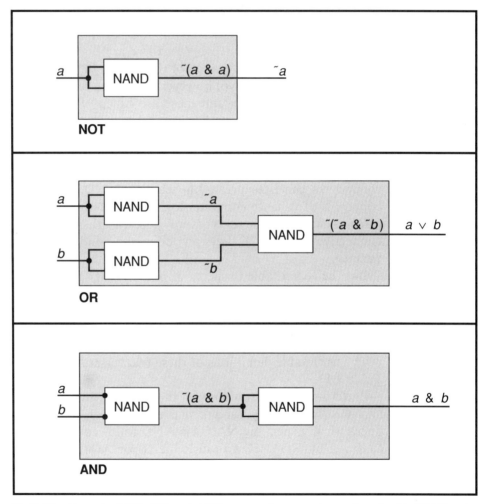

Fig 3.10 How the NOT, OR and AND connectives can be replaced with NAND-only functions

80

How is it the case, however, that each is a basis on its own? We demonstrate it for NAND, using black-box diagramming in Fig 3.10, by showing that AND, OR and NOT can each be built using only NAND boxes. The annotations show how the algebra works.

Thus if we know we can build any logic function using AND, OR and NOT, it follows that we can also do so using only NAND. There may be a greater number of boxes required to achieve it, but that is not always the case in actual practice—much depending on the nature of the function required.

3.5 Tautology and contradiction

Suppose we have set out to model some logical operation by setting up a truth table and putting in bit-values for each row, as we did in the example shown in Table 3.7 in Section 3.1.5. What if every entry were to turn out to be 0? We would have no minterms by which to set up an expression. What would this mean? Or what if every entry were to turn out to be 1? Here, we would have every possible minterm in our expression of the model. In Venn diagram terms it would mean that we would see the result as 'nothing' and 'everything' respectively, but still be somewhat mystified as to the meanings of these outcomes.

It is more fruitful perhaps to start from a different point of view in order to get an insight into what is happening here. Can we think of expressions that would give such results when fully evaluated? The answer is to look first of all at the two constants rules:

$$a \,\&\, \tilde{}a = 0 \qquad a \vee \tilde{}a = 1$$

The truth tables for these give us just the results we have posited (see Table 3.14), and suggest that the 'expressions' we have been seeking are simply 0 and 1.

If this seems a too glib way of getting there, consider a third approach. Suppose that, instead of filling in the truth table, we model direct into the algebra and arrive at an expression such as

$$(a \,\&\, \tilde{}a) \vee (b \,\&\, \tilde{}b)$$

Table 3.15 shows that this too reduces to all 0. Does this mean that the expression itself could be replaced by 0? Yes, it can, as the algebra would

Table 3.14 *Evaluations of {a & ˜a} and {a ∨ ˜a}*

a	$\tilde{}a$	$a \,\&\, \tilde{}a$	$a \vee \tilde{}a$
1	0	0	1
0	1	0	1

Table 3.15 *Evaluation of $(a \& \tilde{~}a) \vee (b \& \tilde{~}b)$*

a	b	$\tilde{~}a$	$a \& \tilde{~}a$	$\tilde{~}b$	$b \& \tilde{~}b$	$(a \& \tilde{~}a) \vee (b \& \tilde{~}b)$
1	1	0	0	0	0	0
1	0	0	0	1	0	0
0	1	1	0	0	0	0
0	0	1	0	1	0	0

confirm

$$(a \& \tilde{~}a) \vee (b \& \tilde{~}b)$$
$$= 0 \vee 0$$
$$= 0$$

The expression is always FALSE no matter what the values of the variables. We could demonstrate similar cases for which the expression is always TRUE. If we designed a switch circuit on the basis of such expressions (before noticing that they would simplify to a constant) we should find ourselves with circuits which could never be switched on (or off) no matter how much we juggled with the individual switches! Logic circuit designers would find themselves with a permanent dead or permanent live output signal no matter what signals the input lines were providing. The Venn diagram results referred to earlier suggest interpretations which seem to be equally non-useful!

However, when we turn to propositional interpretations, the meaning and usefulness of such results become more apparent. For example, a compound logical statement which is 'always true no matter what' is certainly useful—a sound argument should surely have this quality. Finding that some stated logic is always false no matter what, suggests the detection of fallacious reasoning—again a useful result if it is material to some important decision making.

The results are important enough in fact to be given special names: an always-true statement is called a *tautology* and an always-false one a *contradiction*.

The normal forms of expressions can be seen to have some special value in the context of these two results for, when an expression is in conjunctive normal form:

$$c_1 \& c_2 \& \ldots \& c_N$$

only one clause need be 0 for the whole expression to be shown to be a contradiction, and when in disjunctive normal form:

$$c_1 \vee c_2 \vee \ldots \vee c_N$$

only one 'clause' need be 1 for the whole expression to be a tautology. (We put 'clause' in quotes here because the term is conventionally used only in connection with the conjunctive normal form.) It is worth reminding ourselves here that this is because the logical OR is inclusive of the case 'or both true' which is not always how we use the word 'or' in natural language.

EXAMPLE The Sales Manager tells you that The Red Manual is the currently applicable one and must therefore be used. The Marketing Director says that both the Red and the Blue Manuals are current and either can be used. "Well," says the Sales Manager when questioned on this, "Since we both agree that the Red Manual is current there's no contradiction." This is fairly obviously a fallacious argument. How does the logic show it to be so?

Without too much agonising over interpretations, we can say that what we are testing is whether $\{a\}$—Manual A must be used—is equivalent to $\{a \vee b\}$—either Manual can be used.

$$
\begin{aligned}
a &\equiv a \vee b \\
&= (a \,\&\, (a \vee b)) \vee (\tilde{\ }a \,\&\, \tilde{\ }(a \vee b)) \qquad \text{(See 3.3.1)} \\
&= a \vee \tilde{\ }a \qquad\qquad\qquad\qquad\quad\;\, \text{(De-factoring)} \\
&= 0
\end{aligned}
$$

Therefore, there is a contradiction.

"OK," says the Sales Manager, "But what I really meant was that using the Red Manual is not going against what my colleague said."

What does logic say to this? Let us suppose that what we are testing here is the truth of $\{a \text{ IMPLIES } (a \vee b)\}$:

$$
\begin{aligned}
a &\to (a \vee b) \\
&= \tilde{\ }a \vee (a \vee b) \qquad \text{(See 3.3.2)} \\
&= (\tilde{\ }a \vee a) \vee b \qquad \text{(Commutation)} \\
&= 1 \vee b \\
&= 1
\end{aligned}
$$

What we started with was tautologous: the statement $a \to (a \vee b)$ is always true. There is of course the danger that we are still testing the wrong formulation in the first place. But the aim currently is to illustrate results rather than to critically examine the processes of model formulation, so we shall let this pass.

What we have begun to enquire into here is the nature of logical deductive argument, but only in the somewhat passive sense of testing the truth of compound statements. The construction of argument requires rather more subtlety than we have yet to hand, but again we have laid some simple, but firm, foundations.

4 Logical Mapping

4.1 The techniques and tools of systems analysis

Let us first of all sort out some potentially confusing terminology concerning method, technique and so on. Contemporary systems analysis is awash with **methodology**—the study and use of prescribed **methods** for carrying out the work of information systems development and allied tasks. Regarding the types of method available there is frequent use of the prefix 'structured' to convey a message about the internal structure of the method. Even so there are a great number of distinguishable proprietary structured methodologies vying for the attention of the would-be user. While they may share a reasonably common approach to the method structure, their choice of **techniques** (data flow diagrams, entity-relationship diagrams, etc.) may show a mixture of some shared and some 'unique-to-us' offerings. The implementation of the techniques within the methodology will appear as a portfolio of software **tools** with carefully protected trademarked names.

The above paragraph has used a kind of hierarchy of terms regarding ways of doing things, as follows:

Methodology
Method
Technique
Tool.

It is by no means certain that all writers and professionals will use these terms in this way or even in a self-consistent way. There is the additional problem that other terms may crop up in covering the same sort of intended meanings, for example procedure, practice, checklist, step-list, and so on are quite likely to appear in similar roles.

We have carried out this excursion into the methodology question because in this chapter we wish to present some of the representations of logic introduced in the previous chapter as potential techniques for the use of the systems analyst. And we mean techniques in the same sense as the term has been used above. Some of these techniques are beginning to appear as tools associated mainly with the use of expert sysem shells, although some logic tools related to program design have been in existence for over a decade, though never perhaps widely adopted. Most of the techniques have been available for much longer (K-mapping and decision tables, for example), but have rarely found their way formally into all-embracing methodologies for systems analysis. This situation will probably change as knowledge-oriented systems become part of the remit of systems analysts' work.

Systems analysts have always been expected to show commitment to the use of diagramming, charting and various tabular techniques in the performance of their work, in both the analytical and the design aspect of it. These techniques have lately been developed into software-supported tools, and as these products have become increasingly sophisticated they have aimed to portray an image of the analyst at work using an advanced workstation to handle not only the formidable task of chart drafting and modification but also that of maintaining a knowledge encyclopaedia of the modelled data—its names, descriptions, uses, cross-references, and so on. This is a fine 'engineering' image, but there is a danger that it is one that will promote the view of a designer driven by tools rather than one who is a creative user of those tools.

The profession needs to continue to attract people of well-rounded skills, including the social and creative ones, if computer-based systems are to be both functional and attractive, both economic and user-friendly. To this end it is better that the systems analyst be provided with a well-stocked **tool-box** of smart (=clever) tools from which to select 'the right one for the right job', rather than a handle-turned box of tricks which produces machine-designed systems by rigid procedures known only to the machine, however useful this may be as relief from the drudgery of certain types of design work. This chapter offers ideas for extending the range of tools in that tool-box, and not necessarily software-driven tools either—they may be equally effective as paper tools for use on the office desk, or even on the lap during face-to-face communication.

In any case, sharing the view of a chart with another, whether that other be user, colleague or supervisor, is usually many times more rewarding than the use of the spoken or written word on its own. Even a list of bullet-points, simple though the concept of such a 'chart' is, is somehow easier to deal with where minds are meeting than merely being lectured at or being told to "read the report". One person checking that they have understood correctly the meaning of another is, or ought to be, a very common scenario in systems analysis, and diagrams, charts, tables and lists are excellent aids to an eventually satisfactory outcome.

As is the case with already familiar techniques in systems analysis, these logic-related techniques can also be of immense value to the analyst closeted in the office with a mass of information, with a need to sort it out and with an objective of wanting to achieve a clear and structured view of it all. Much of this work may be sketchy and eventually consignable to the WPB, but while it is in progress the brainwork needs all the assistance it can get. The reader is asked to bear all this in mind while reading this chapter. It should make a difference to understanding its significance, and the examples have been designed with that thought in mind too.

As a bridge between this and the previous chapter, we shall start with K-maps as a technique for rapid simplification of logical expressions, to show that to some extent the algebra *can* be bypassed. But this should also lead to the notion that K-maps have merit in their own right as paper tools. Venn diagrams are referred to in this context very much as less-formalised K-maps, though their use as illustrators of classes, categories and sets will not be forgotten, especially in relation to the next chapter, which is on this very topic area.

Decision tables will then be given a thorough treatment to show how they fit into the gap between logic tree diagrams and the pure formality of truth tables and K-maps. They also provide a useful common-sense insight into the question of what lies beyond two-valued systems.

Finally the least-standardised area of logic charting will be looked at with a view to preparation for the more difficult knowledge processing ideas of Chapter 6.

4.2 K-maps as analytical techniques

4.2.1 Introduction

First we examine thoroughly the significance of the structural form behind the K-map. From one viewpoint it is simply a re-orientation of a truth table, each cell of the map corresponding to the bit-value entry in the result column of the table. From another viewpoint it is a standardised Venn diagram, where each logical sub-area of such a diagram has been squeezed into the standard square shape of the map cell. This correspondence is illustrated by Tables 4.1(i) and (ii) and Fig 4.1, which show {$a \lor b$} in the three forms—table, map and diagram—appropriately labelled to emphasise the equivalent meanings.

The truth table when fully set out for an evaluation exercise is more explicit than the K-map in showing the step-by-step working. The Venn diagram has advantages of illustrative powers in showing the variables as easily recognisable circles with characteristically shaped sub-areas for the

Table 4.1 *Truth table and K-map for $a \lor b$*

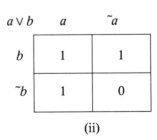

a	b	$a \lor b$	Minterm
1	1	1	$a \& b$
1	0	1	$a \& \tilde{}b$
0	1	1	$\tilde{}a \& b$
0	0	0	—

(i)

$a \lor b$	a	$\tilde{}a$
b	1	1
$\tilde{}b$	1	0

(ii)

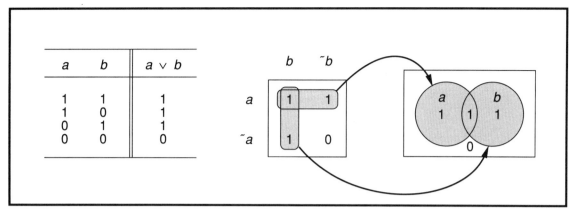

Fig 4.1 Showing the correspondence between truth table, K-map and Venn diagram for the expression $a \vee b$

various shared and non-shared sectors, which the K-map makes less visible with its standardised cells. However the K-map has a very distinct advantage over both these other forms when it comes to recognising directly from it the simplest form for the algebraic expressions; it permits this without use of the minterms and with very little algebraic manipulation.

4.2.2 K-maps for three and four variables

To appreciate fully this special power of the K-map it is desirable first to consider how we set up such maps for expressions in more than two variables. Since K-maps are essentially coordinate systems, three variables would seem to imply the need for three coordinate dimensions and therefore for three-dimensional diagrams. This would indeed be feasible and even desirable in one sense, but since we are looking to go above this number of variables and yet stay with a two-dimensional medium (the paper surface) we need a less-demanding technique. The right effect is achieved by doubling-up the coordinate dimensions. This is shown in Fig 4.2 for both three and four variables. The number of cells is 2 to the power of 3 ($=8$) and to the power of 4 ($=16$) respectively, as we would expect. Each cell is uniquely identifiable by coordinating the variables as before.

There are two important points about the labelling which should be noted immediately. One is that neighbouring labels must never differ by more than one negation change at a time ($a \& b$ next to $\tilde{~}a \& \tilde{~}b$ would not be correct). We can trace the reason for this back to the Venn diagram, where the crossing of a boundary between sub-areas can only ever concern the move into or out of one of the variables circles, thus reversing the negation state of that one variable. Fig 4.3 illustrates this. (Crossing at junctions would not be the same as lateral or vertical crossings in the K-map.)

The second point is that, so long as this constraint is observed, it does not matter which label we start with in the sequence (across or down) or how

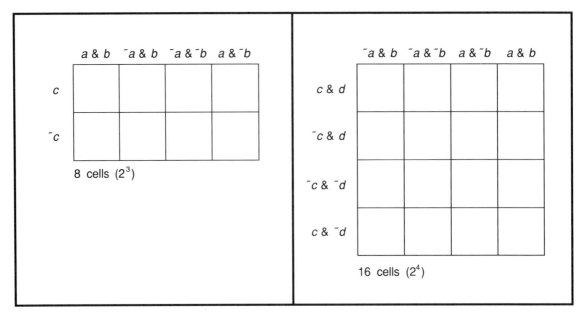

Fig 4.2   The K-map frames for three and four variables

the sequence then proceeds, since there are many ways of mapping from the Venn diagram into the K-map. This is illustrated in Fig 4.2 by labelling the coordinates differently in (i) and (ii). The constraint is observed while the above freedoms are made use of.

We are now ready to observe that within the K-maps there are certain inner shapes, known as **sub-cubes** (because of their generality of dimension). These sub-cubes can in most cases be labelled using fewer than

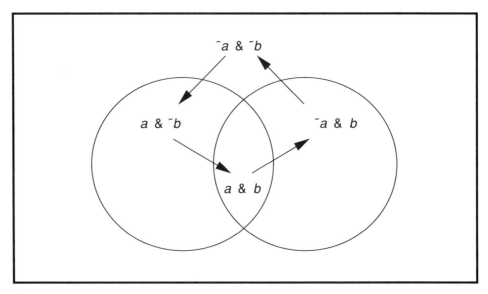

Fig 4.3   The single negation change from cell to cell

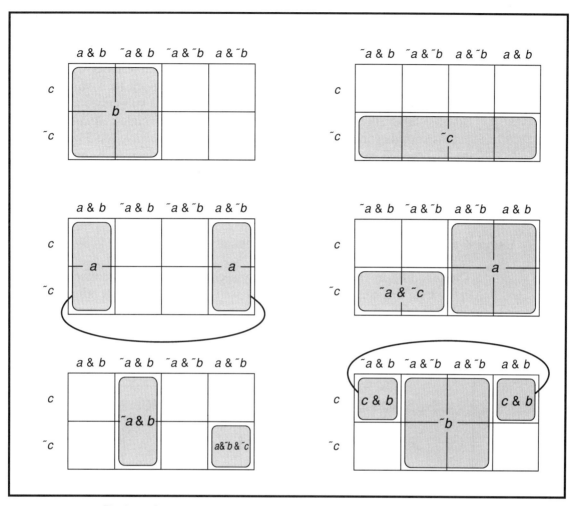

Fig 4.4 Some sample simplification sub-cubes on a three-variable K-map

the full number of variables, ie. with less complexity than the minterm labels. There are a random collection of such observed sub-cubes in Fig 4.4. For the sub-cubes with a single-variable label, we are merely observing the transformed circle of the Venn diagram; for two-variable labels we are observing the overlaps between two circles, and so on.

In fact we can generalise about what sub-cubes we should be able to observe as follows. A sub-cube comprises groups of adjacent cells forming any complete rectangle, the number of cells always being a power of two, ie. 2 or 4 or 8. Because of the arbitrariness of the labelling, 'adjacency' applies at the edges of the map also!

There is also a relationship between the number of cells in a sub-cube and the simplicity of the label. In a two-variable map a single-variable label, eg. a, $\tilde{a}$, b, $\tilde{b}$, will apply to a two-cell sub-cube and a two-variable label, eg. $a \& b$, $\tilde{a} \& b$, will apply to a single-cell sub-cube—the minterm in fact as we would expect. In a three-variable map a single-variable label will apply to a four-cell sub-cube, a two-variable label will apply to a two-cell sub-cube, and a three-variable label (the minterm again) will apply to a single-cell sub-cube. In a four-variable map, a single-variable label will apply to an eight-cell sub-cube, a two-variable label will apply to a four-cell sub-cube, a three-variable label will apply to a two-cell sub-cube, and a four-variable label (the minterm again) will apply to a single-cell sub-cube.

This sounds complicated when written out in this way, but Fig 4.4 shows the facts to be correct, and the meaning of the pattern of numbers is that a single variable must occupy half the map (itself and its negation together cover the whole map because $\{a \lor {\sim}a = 1\}$), a two-variable term must occupy a quarter of the map and so on.

4.2.3 Using K-maps for simplification of expressions

The labelling in the last section may seem to be rather inconsequential at first, but the act of labelling is itself a substitute for the algebraic simplification of expressions and, once mastered, much easier to get right. The reduction in the number of stages is described in Fig 4.5.

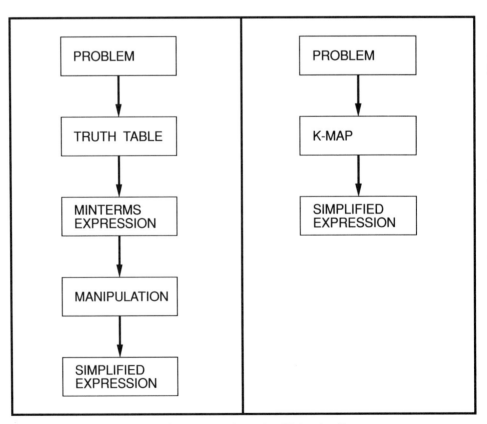

Fig 4.5 The fewer stages of working when simplifying by K-map

But what exactly is the problem-to-map stage? Let us reintroduce the example which was worked through in Chapter 3. A truth table was drawn up in Section 3.1.6 and the algebraic simplification carried out later in Section 3.2.5. The example was:

EXAMPLE   An alarm circuit is required which can be controlled from the use of three switches. One switch is to be a master switch, switching the circuit on and off whatever the state of the other two. Otherwise the circuit should be switched on if any two (or all three) of the switches are on.

We shall again label the three switches a, b, and c, with c being assigned the role of master switch and now, rather than trying to interpret the example as an algebraic expression direct, we draw up the K-map as shown in Fig 4.6(i). Remember that ones and zeros represent the circuit being ON and OFF, while the state of the switches is represented by the variable being in negated form or not. Once again we insert the bits in the Alarm circuit map, carefully interpreting the requirements described. Thus whenever c is not negated, the circuit is 1. Whenever two out of three of a, b, or c are not negated, the circuit is 1.

Now we seek out the simplifying sub-cubes. The larger the sub-cube, the simpler the label, and therefore we maximise the opportunities by re-using cells if need be. This is entirely feasible since it is only the fact that the cell represents TRUE that is relevant, not how many times it does so. Each labelled sub-cube is in itself TRUE ($=1$) and therefore represents a clause in a disjunctive normal form. We can therefore write out straight from the map the fully simplified expression for the alarm circuit:

$c \lor (a \& b)$

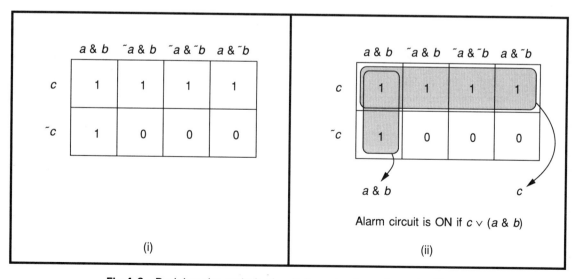

Fig 4.6   Deriving the switch circuit formula direct from the K-map

Compare this with the algebraic simplification we carried out in Section 3.2.5!

4.2.4 The use of don't-care values

When specifying the logic conditions needed to satisfy some problem situation there can sometimes be cases where some combinations of states of the constituent variables are immaterial to the overall function or purpose. K-maps show very explicitly how such don't-care values can be taken great advantage of in achieving maximum simplification.

For example, suppose in the alarm circuit example there had been a fourth switch (which we shall label *d*), whose presence though necessary for some technical reason in other uses of the circuitry is immaterial to the alarm functions we have described. If we were to interpret this as meaning that it can be left switched off we would arrive at the K-map shown in Fig 4.7.

The simplification from this is

$$c \vee (a \& b \& \tilde{\ }d)$$

as shown. But if we include the don't-care cell as shown in Fig 4.8, the simplification can remain as

$$c \vee (a \& b)$$

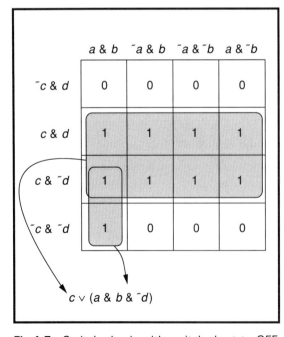

Fig 4.7   Switch circuit with switch *d* set to OFF

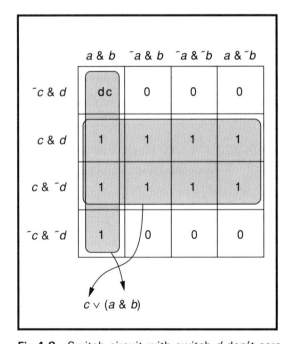

Fig 4.8   Switch circuit with switch *d* don't-care

4.3 Decision tables as analytical techniques

4.3.1 Introduction

As we have seen, truth tables and K-maps provide useful alternative methods for setting out systematically all possible values of a Boolean function. If we were to regard each of the variables as representing a binary decision option and the function as representing a decision consequent upon these, then we have a rudimentary method for representing or modelling a simple decision structure. This was introduced in Chapter 2.

Take for example a clerk processing customer orders using a known procedure (or decision structure) which includes the following:

> Approve and accept the Order if the Customer's credit limit will not be exceeded by this Order and there is no Restriction on the goods ordered. If the Restriction has been given special clearance however the Order can still be accepted. If the credit limit would be exceeded only approve the Order if this Customer's paying record is good, though again special clearance can be used to overcome this.

Let us define five Boolean variables which would appear to cover the above contingencies as follows:

A Credit limit exceeded?
B Restriction on goods?
C Paying record good?
D Special clearance given?
E Approve Order?

By presenting the variables as questions and adopting the convention that 1 represents the answer 'yes' and 0 the answer 'no', we could treat E as a function of the variables A, B, C ,D and draw up a truth table as shown in Table 4.2.

What if anything do we gain by doing this? One advantage is that in order to arrive at the truth table we have had to carry out a very careful analysis of the narrative instructions. You should get the flavour of this by selecting two or three lines of the table and checking for yourself that in each case the value of the function variable, E, is what you would expect. (We have had to make one important assumption of course, that of the absence of E = yes in the narrative, ie. not specifically stating that the order can be approved we have taken to imply that E = no.)

The justifications for the E = 1 entries are as follows. Whenever D = 1, then E = 1 (special clearance). When C = 1 with A = 1, then E = 1 (good paying record). When both A = 0 and B = 0, then E = 1 (straightforward acceptance).

So the truth table as a tool for analysing decision structure could prove very useful. And yet when we look at it as a final representation of that

Table 4.2 *Truth table modelling for the Order Processing example*

A	B	C	D	E
1	1	1	1	1
1	1	1	0	1
1	1	0	1	1
1	1	0	0	0
1	0	1	1	1
1	0	1	0	1
1	0	0	1	1
1	0	0	0	0
0	1	1	1	1
0	1	1	0	0
0	1	0	1	1
0	1	0	0	0
0	0	1	1	1
0	0	1	0	1
0	0	0	1	1
0	0	0	0	1

analysis we could hardly claim that it was clear and readable. Even if we were to use the more condensed K-map form (perhaps in order to explore the best algebraic simplification) we would still find it difficult to move between the narrative and the logic representation with any great ease.

Part of the problem stems from our having resorted to symbolic representation of the variables. Would the table be more acceptable if we write out the questions as column headings? Possibly, but then we have a somewhat clumsy-looking table, heavy with headings and thin in entries. This might look more balanced if we write our yes and no entries out in full.

Another disadvantage is not actually revealed by the example we have taken, simply because the instructions were all directed towards one function, E: Accept Order?, and this is not at all typical of such instructions in practice. Let us now add some further sentences to make the example more realistic. We shall assume that the instructions continue with

> Where special clearance has not be given for an Order which involves both a Restriction and an exceeding of the credit limit, refer it to the Chief Clerk. Orders for which the credit limit is not exceeded but are from Customers who do not have a satisfactory paying record and for whom no special clearance has been obtained should be returned to Sales Department.

Effectively there are now two more functions which we could identify as

F   Refer to Chief Clerk?
G   Return to Sales Dept?

Table 4.3 *Order Processing: the multi-functional truth table*

A	B	C	D	E	F	G
1	1	1	1	1	0	0
1	1	1	0	1	1	0
1	1	0	1	1	0	0
1	1	0	0	0	1	0
1	0	1	1	1	0	0
1	0	1	0	1	0	0
1	0	0	1	1	0	0
1	0	0	0	0	0	0
0	1	1	1	1	0	0
0	1	1	0	0	0	0
0	1	0	1	1	0	0
0	1	0	0	0	0	1
0	0	1	1	1	0	0
0	0	1	0	1	0	0
0	0	0	1	1	0	0
0	0	0	0	1	0	1

We now interpret $F = 1$, when $A = 1$, $B = 1$ and $D = 0$. The case $G = 1$ is when $A = 0$, $C = 0$ and $D = 0$.

The truth table now needs to be *multi-functional* (not at all suited to a single K-map representation, by the way) and would appear as shown in Table 4.3.

The table is now even wider and attempting to represent it with all column headings written out in full will be even more unsatisfactory. However the table does still have useful analytical properties. We could observe for example that the functions F and G are mutually exclusive since only one yes appears for them in any row. Such 'tidiness' is not what we may always expect to find in actual cases.

We may also notice that the eighth and tenth rows both contain all zeros in the function columns, which means that they have no defined action under our current interpretation of the procedures. This should direct the analyst to raising the question with an appropriate authority to verify whether this is acceptable or not. We shall in the later Section 4.4 assume that this has been done, but for now we leave these condition sets without any defined action attached to them.

However, there is a different, and some would say better, technique for setting out on such an analysis. This is the Decision Table format which we introduced briefly in Chapter 2.

4.3.2 The decision table format

The decision table format overcomes the column heading problem in a very simple and obvious way—it lays the table on its side so that the column headings become row headings! Of course the question headings will still take up a wide space but now there will be only one question per row and the remainder of entries are the Boolean values. Also we prefer to read the decision conditions before we read the corresponding action or actions and therefore we lay the variables rows above the functions rows. Finally we compromise on the value entries by using Y and N in place of either 1 and 0 or the full words. The table from the previous section now looks as shown in Table 4.4.

Notice some other features of this format. First, we have retained the symbolic representations of the questions for ease of reference to the table itself; 'see Row D' is a convenient and unambiguous instruction. Second, we have dropped the question marks since they can be taken as read, since the responses are either Yes or No. And third, we have adopted a different method of flagging the function values to emphasise the 'do this' nature of this part of the table, namely by inserting X where action is indicated.

Another useful concept that has been introduced at this point is that of the **rule**. A rule in a decision table is any individual and complete column of entries since it can be read as a self-contained instruction as to what action to take under specified conditions. It is therefore useful to introduce the column reference numbers (or rule numbers) so that specific rules can be cited precisely.

A decision table is now seen to comprise the following main components. An *upper half of condition rows*, each comprising a title (or *stub* as it is often called) and a set of condition values or *entries*, and a *lower half of action rows*, each comprising a stub and a set of action entries. Divided vertically we see that the table comprises a *left half of stubs*, appropriately referenced, and a *right half of entries* formed into columns of rules, appropriately referenced. It is useful to emphasise this four-quadrant structure by using double or heavy lines to separate the quadrants.

Table 4.4   *The decision table re-layout of Table 4.3*

	0									1						
Rule	1	2	3	4	5	6	7	8	9	0	1	2	3	4	5	6
A: Credit limit exceeded	Y	Y	Y	Y	Y	Y	Y	Y	N	N	N	N	N	N	N	N
B: Restriction on goods	Y	Y	Y	Y	N	N	N	N	Y	Y	Y	Y	N	N	N	N
C: Paying record good	Y	Y	N	N	Y	Y	N	N	Y	Y	N	N	Y	Y	N	N
D: Special clearance	Y	N	Y	N	Y	N	Y	N	Y	N	Y	N	Y	N	Y	N
E: Approve Order	X	X	X	–	X	X	X	–	X	–	X	–	X	X	X	X
F: Refer to Chief Clerk	–	X	–	X	–	–	–	–	–	–	–	–	–	–	–	–
G: Refer to Sales Dept	–	–	–	–	–	–	–	–	–	–	–	X	–	–	–	X

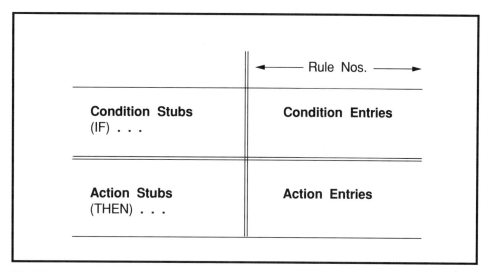

Fig 4.9 The four quadrant areas of a decision table

Figure 4.9 illustrates these features diagrammatically, with the IF...THEN syntax appended to remind us of the basic purpose of the format.

Although there have without doubt been many users of the decision table technique who have been (perhaps blissfully) ignorant of its Boolean ancestry, it is valuable for a logical analyst to keep that ancestry in mind because, although the format is user-friendly, it has rather disguised the rigour of the logic that it is intended to model.

4.3.3 Consolidation of tables

The practised eye of the logician will have spotted immediately that the example in the previous section contains redundancy—there is potential for reduction through simplification. There is a complication in that the table deals simultaneously with three actions (E, F and G) and, if we regard each combinational pattern of actions as a different function, there are five functions (E, F and G alone, and also the combinations E with F and E with G). The simplification could still be tackled using algebra or K-mapping, but for the less mathematically confident systems analyst the task could prove daunting, especially if more actions, and functions arising from their combination, are involved.

A procedure known variously as *consolidation*, reduction or abbreviation (we shall use the first of these) has therefore been devised for use with the decision table format, which turns the simplification procedure into a repeated simple-step process, which though it can be tedious to apply manually is less mathematically demanding. It is actually based on the defactoring of disjunctive normal form clauses which we can generalise as follows:

$$(a \,\&\, b \,\&\, c \ldots \,\&\, n) \lor (\tilde{\,}a \,\&\, b \,\&\, c \ldots \,\&\, n)$$
$$= (a \lor \tilde{\,}a) \,\&\, (b \,\&\, c \ldots \,\&\, n)$$
$$= 1 \,\&\, (b \,\&\, c \ldots \,\&\, n)$$
$$= b \,\&\, c \ldots \,\&\, n$$

The significance of this simplification is that the two original clauses were identical apart from a negation of one of the variables. The variable in question can thus be eliminated as shown, leaving just one clause.

Now, since a decision table is, in terms of the logic, a lateral truth table, it follows that each condition portion of a rule column is in fact a clause. Rules which lead to the same action-pattern belong to the same Boolean function and are therefore open to the same simplification as above if they satisfy the above condition in terms of differing in the state of only one variable. But the reduction from two clauses to one can now be stated in the much more practical-seeming terms of consolidating two rules into one! The eliminated condition does not however vanish from the scene because it is, in general, needed for other rules in the table. We introduce the don't-care to indicate that the variable has been eliminated for the rule in whose column it appears. (Don't-care is usually presented as a dash in decision table conventions.)

The consolidation of two rules is illustrated in the schematised view of a portion of a decision table in Fig 4.10. The common-sense interpretation is that, since the outcome of condition a is immaterial to taking the action, then we don't care what state it is in for this rule.

The **consolidation procedure** can be more formally stated as follows:

> If two rules leading to the same action in a decision table have identical condition patterns except in respect of one condition entry, which shows a Y/N opposition, then they can be consolidated into a single

Rule No.		1	2		Rule No.		1′
Condition	a	Y	N			a	-
	b	N	N	*consolidates to*		b	N
	c	Y	Y			c	Y
	d	Y	Y			d	Y
Action		X	X		Action		X

Fig 4.10 The consolidation of two rules into one

rule showing a don't-care entry in the position of the originally differing condition entries.

Although this procedure may appear a little clumsy to the mathematically minded it has the virtue that it can be applied 'mechanically' and repeatedly, the repetition often taking previously consolidated rules and further consolidating them.

The algebra required to achieve this with even modest numbers of variables is far from straightforward and the K-map approach could require the construction of maps for numbers of variables greater than four, which we have not dealt with (though it requires only a modest amount of ingenuity to set up such maps on a similar basis to that already adopted for three and four variables).

We shall illustrate repeated consolidation by working through the table arrived at in the previous section. However, we shall assume that the queries mentioned in Section 4.3.1 regarding rules 8 and 10 have been attended to and that the correct action for these has in both cases been confirmed to be Refer to Chief Clerk (see Tables 4.4 and 4.5).

Remembering that we can only consolidate rules leading to the same actions (ie. simplify clauses for the same function), we identify candidate rule-pairs which differ in only one condition-entry as follows:

	Common action	Differ at row
Rules 1 and 3	E	third
Rules 4 and 8	F	second
Rules 5 and 6	E	fourth
Rules 7 and 15	E	first
Rules 9 and 11	E	third
Rules 13 and 14	E	fourth

Note that, although there are other possible pairings (eg. Rules 1 and 5), a rule can only be used once in consolidation. The resultant rules may

Table 4.5 *Table 4.4 with rules 8 and 10 updated*

	0									1						
Rule	1	2	3	4	5	6	7	8	9	0	1	2	3	4	5	6
A: Credit limit exceeded	Y	Y	Y	Y	Y	Y	Y	Y	N	N	N	N	N	N	N	N
B: Restriction on goods	Y	Y	Y	Y	N	N	N	N	Y	Y	Y	Y	N	N	N	N
C: Paying record good	Y	Y	N	N	Y	Y	N	N	Y	Y	N	N	Y	Y	N	N
D: Special clearance	Y	N	Y	N	Y	N	Y	N	Y	N	Y	N	Y	N	Y	N
E: Approve Order	X	X	X	–	X	X	X	–	X	–	X	–	X	X	X	X
F: Refer to Chief Clerk	–	X	–	X	–	–	–	X	–	X	–	–	–	–	–	–
G: Refer to Sales Dept	–	–	–	–	–	–	–	–	–	–	–	X	–	–	–	X

Table 4.6 *Table 4.5 after first-round consolidation*

	0						1			
Rule	1'	2	4'	5'	7'	9'	0	2	3'	6
A: Credit limit exceeded	Y	Y	Y	Y	–	N	N	N	N	N
B: Restriction on goods	Y	Y	–	N	N	Y	Y	Y	N	N
C: Paying record good	–	Y	N	Y	N	–	Y	N	Y	N
D: Special clearance	Y	N	N	–	Y	Y	N	N	–	N
E: Approve Order	X	X	–	X	X	X	–	–	X	X
F: Refer to Chief Clerk	–	X	X	–	–	–	X	–	–	–
G: Refer to Sales Dept	–	–	–	–	–	–	–	X	–	X

however be used as candidates for further consolidation as we shall see below. So the table can now be consolidated down to that shown in Table 4.6.

But now there are further consolidations possible. Note that we must be careful to match don't-cares in establishing at which entries rules do not differ. We cannot say that the dash does not differ from Y (or N) because it is a don't care. (The don't-care here is a more specific type than we met in K-map simplification because we prefer to avoid ending up with two consolidated rules sharing a common non-consolidated rule; see Chapter 6 discussion of rule overlap.)

Rules 1' and 9' consolidate to Rule 1".
Rules 5' and 13' consolidate to Rule 5"

The table now looks as shown in Table 4.7.

No further consolidation is possible with this table. Rules 12 and 16 look tempting. They differ in only the second row and share an action. But the action patterns are not identical and therefore the rules belong to different

Table 4.7 *Table 4.5 after second-round consolidation*

	0					1		
Rule	1"	2	4'	5"	7'	0	2	6
A: Credit limit exceeded	–	Y	Y	–	–	N	N	N
B: Restriction on goods	Y	Y	–	N	N	Y	Y	N
C: Paying record good	–	Y	N	Y	N	Y	N	N
D: Special clearance	Y	N	N	–	Y	N	N	N
E: Approve Order	X	X	–	X	X	–	–	X
F: Refer to Chief Clerk	–	X	X	–	–	X	–	–
G: Refer to Sales Dept	–	–	–	–	–	–	X	X

functions and cannot be taken as candidates. If we were to draw up a table for Action G only, it would be a different matter of course because the actions would be the same for the two rules. This is a useful first insight into the question of breadth of scope of analysis. It might well be worthwhile narrowing the scope, ie. drawing up the Action-G-only table, in order to check with Sales under what conditions they themselves believe orders are referred to them, for example. It is not unusual to find that this approach highlights differences in understanding of procedures between two individuals or departments.

In Chapter 6 we shall see how there are more ways in which these tables can be used to check the analysis of procedures for consistency and completeness.

4.4 Tableaux for logic

4.4.1 Introduction

We have looked at the derivation of decision tables as specially oriented and labelled truth tables. But decision tables have long held a place in the system analyst's tool-kit without there being necessarily an awareness of that derivation. This suggests that there is something of intrinsic value in the decision table which is perhaps worth bringing out. Part of the answer lies in the layout conventions that were adopted. (Another part lies in the value accruing from being able to carry out completeness and consistency checking, which we deal with in Chapter 6.)

Whether fully aware of the underlying logic principles or not, the adept user of a decision table is accepting that the rule-column configuration represents valid alternative conditions-to-actions (ie. disjunctive OR connectivity) comprising necessary combinations of the listed condition-values (ie. conjunctive AND connectivity). Thus the layout itself is an integral part of the logical syntax. We now ask whether this syntax cannot be exploited further and whether there are alternatives to such an approach.

We generalise this approach by referring to such layouts as *logical tableaux*.

4.4.2 Nested tableaux

The decision table is divided vertically into the right-hand quadrants containing the *entries* (Y, N, X and −) and the left-hand quadrants displaying the row-labels (or *stubs*) which interpret the meanings of the entries. This layout is designed to achieve a balance between readability and succinctness, and this is fine in so far as its use as an analytical technique goes. However, when the systems analyst comes to discuss the accuracy of the table and its implications with a non-versed user there may be troublesome difficulty in the ability to interpret and 'read back' into everyday understanding. This is in no little part due to the labels/entries separation.

If, however, suitably recognisable abbreviations for the labels can be devised, they can be brought back into the main logic structure in place of the single-character entries, and a tableau of the logic can be created which uses the lateral disjunction and vertical conjunction to display, in nested boxes, the essential meanings of the rules and rule sets.

This has been carried out for the Approve Order action in our order processing example and shown as Fig 4.11. The rules have been re-shuffled with respect to their original left-to-right sequence in Table 4.7, partly for general readability purposes but also because two rules are shared with the other two actions, and this leaves the way open to display the three action envelopes overlapped in a Venn-diagram approach to the illustration as shown in Fig 4.12.

There are various alternatives that could be adopted, appropriate to the user-friendliness aspect of the work. For example, the boxes around the

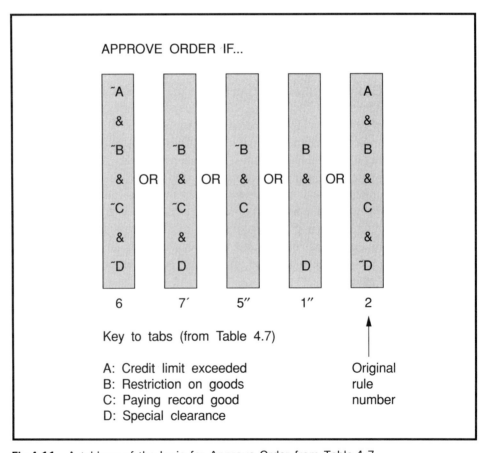

Fig 4.11 A tableau of the logic for Approve Order from Table 4.7

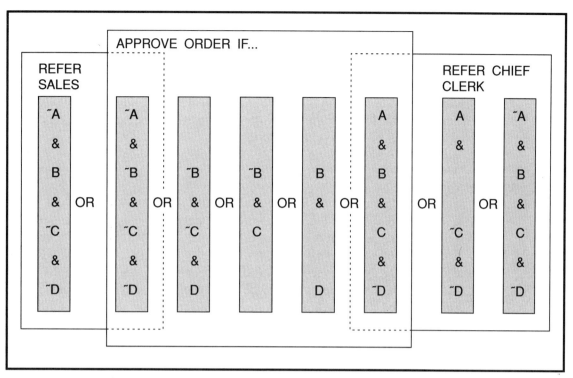

Fig 4.12 Overlapping tableaux for the whole of Table 4.7

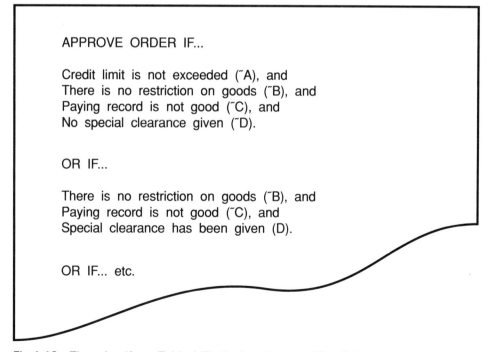

APPROVE ORDER IF...

Credit limit is not exceeded (˜A), and
There is no restriction on goods (˜B), and
Paying record is not good (˜C), and
No special clearance given (˜D).

OR IF...

There is no restriction on goods (˜B), and
Paying record is not good (˜C), and
Special clearance has been given (D).

OR IF... etc.

Fig 4.13 The rules (from Table 4.7) displayed as condition lists

conditions could be dropped and the action titles brought over the consequent lists as column headings (see Fig 4.13). This brings us closer to the kind of listing one might well sketch out when analysing the original source of information for the analysis of the procedures (see Chapter 6), but here we know that the listing is optimal, having derived it from a fully worked decision table, whereas notes taken during analysis are likely still to contain at least some of the incompletenesses, redundancies and contradictions of the source or sources.

4.4.3 Slabbed tableaux

It may prove useful to highlight how, where there are several sets of alternatives for an action such as is the case for Approve Order, certain conditions are shared. This is shown in Fig 4.14 for the conditions CR OK and CR not OK, our chosen abbreviations for the original condition: Credit limit exceeded. (Note that this device also reversed the truth values!). This *slabbing* across the diagram helps to focus attention on the credit approval aspects of the procedures should one wish to do this in discussion with the user. The shared condition sets are in this case the don't-care items and could if necessary be omitted.

Full slabbing would of course be achievable only by going back to the unsimplified decision table, where all condition values are set out explicitly and in blocked sets. This draws attention to the necessary hierarchy that the sequence of listing of the conditions brings with it (see Fig 4.15). There is a good deal of practical importance which attaches to the sequencing of condition lists and action lists and to rule ordering which we shall address in Chapter 6.

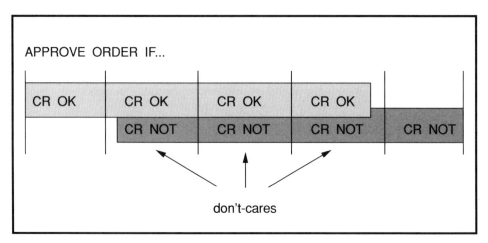

Fig 4.14 Slabbing of a condition to show up how it occurs in the rules

	Rule	0									1						
		1	2	3	4	5	6	7	8	9	0	1	2	3	4	5	6
A: Credit limit exceeded		Y	Y	Y	Y	Y	Y	Y	Y	N	N	N	N	N	N	N	N
B: Restriction on goods		Y	Y	Y	Y	N	N	N	N	Y	Y	Y	Y	N	N	N	N
C: Paying record good		Y	Y	N	N	Y	Y	N	N	Y	Y	N	N	Y	Y	N	N
D: Special clearance		Y	N	Y	N	Y	N	Y	N	Y	N	Y	N	Y	N	Y	N
E: Approve Order		X	X	X	-	X	X	X	-	X	-	X	-	X	X	X	X
F: Refer to Chief Clerk		-	X	-	X	-	-	-	X	-	X	-	-	-	-	-	-
G: Refer to Sales Dept.		-	-	-	-	-	-	-	-	-	-	-	-	X	-	-	X

Fig 4.15 Full slabbing of the conditions in Table 4.5

4.5 Tree-structured logic charts

4.5.1 Introduction

Tree structures play an enormously important role in computer science. Not only do they provide, in diagrammatic form, graphic interpretations of search-path alternatives, hierarchical classification of data and procedures, and of well-structured software components, but they can even exist in coded form as data which drives and controls program action. These technical matters can be of interest to the systems analyst concerned with good-quality software design and we shall have something to say about this in Chapter 6 and elsewhere.

But our concern here is with the tree structure as a basis for the charting of logic and consequently as a possible technique for use by the systems analyst directly. There are three contrasting approaches to using a tree structure to map logic. One is the decision-tree approach which we have already met and which follows naturally from the discussion of trees and tableaux in the previous section. The second is indentation, an adaptation of the horizontal tree structure, which forces the logic to be read in a given sequence. This is of especial importance to the mapping of traditional 'manual' programming logic (prior to fourth generation language programming) since the coded program statements have to be stored as a listing which defines the sequence in which the instructions are to be interpreted and turned into computer action.

The third approach is to directly represent connectives in some diagrammatic form suited to depicting the hierarchic interpretation of the logic. We shall confine our interest here to the AND-OR mapping approach which plots the structure of Boolean expressions and can therefore be useful as a device for looking at the 'shape' of expressions in a visual way. To a certain extent it can also be used as an alternative to algebraic manipulation and without the limitation of number-of-variables that we have encountered with truth tables and K-maps.

4.5.2 Binary decision trees

We can understand something of the basic nature of the decision tree by considering it as a direct translation of the slabbed tableau in Fig 4.15. The slabbing has already visually given a sense of hierarchy to the condition set and it is this hierarchy which the tree displays quite literally (see Fig 4.16).

Since Fig 4.15 displayed the tableau of the equivalent of a full truth table, so the tree in Fig 4.16 displays every possible combination of condition tests and there is little to choose between them. They both give a potentially useful, if somewhat space-hungry, view of the detail of the truth table.

In practice we are much more likely to make use of the tree as a means of displaying the decision structures that are in actual use in order to check our understanding of them and to help to establish with users the accuracy of our understanding. Such trees will be very far from being a full display of every condition-test combination. Reflecting actual human usage they are likely to incorporate pragmatic consolidation and even some redundancy, and will also not display at all any test combinations that are never used or make no sense in the real world context. They may also contain hidden contradictions which may never have come to light in actual practice or, if they have, have been treated as anomalies or exceptions and left in situ.

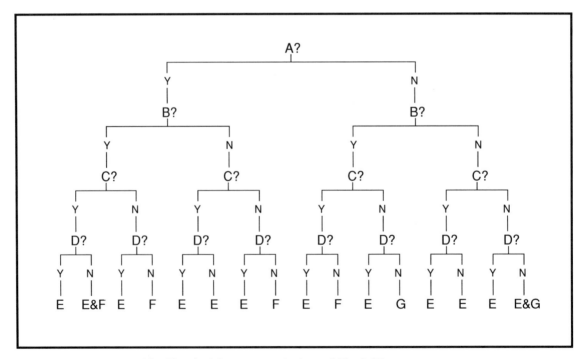

Fig 4.16 The decision tree equivalent of Fig 4.15

4.5.3 Indentation

It is a relatively trivial transformation to rotate a tree structure until its trunk-to-branch direction is horizontal rather than vertical (see Fig 4.17(i)). Indeed there is little to choose between the two orientations, since they both suit the left-to-right, top-to-bottom reading mode of Western cultures. However, the horizontal orientation is closer to the indentational format as Fig 4.17(ii) shows and this why we have used it as an intermediate illustration. The indentational format is an explicit imitation of formal report style, where indentation is used to indicate that a paragraph is subordinate to another paragraph. Narrative is also, of course, intended to be read in a line-by-line sequence and the indentational format is likewise

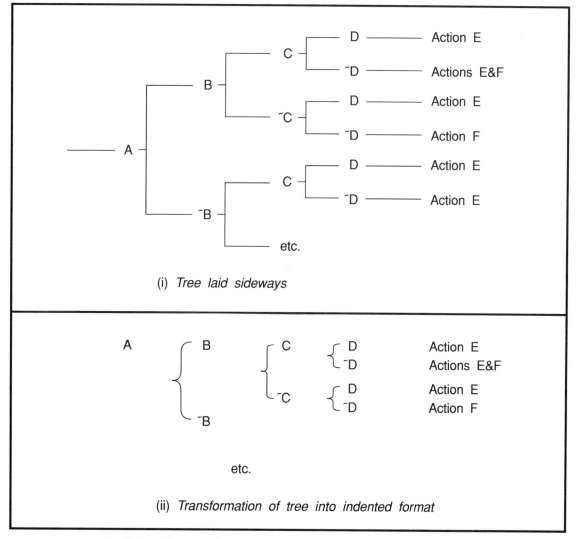

(i) *Tree laid sideways*

(ii) *Transformation of tree into indented format*

Fig 4.17 Showing the connection between tree and indented forms

intended to be read in this way. Whereas a tree invites the reader to survey its contents in any number of ways and to interpret the topography of the branches and sub-branches in a somewhat arbitrary way (unless specific conventions are in force, of course), the indented format is inevitably to be taken as meaning that the sequence of reading is fixed.

The Action Diagram conventions of Martin and McClure and the Structured Diagrams of Warnier-Orr propose a visual syntax for such an indented structure. The chief aim is to provide a tool for clearly structured program design. The basic idea is a simple one. Successive depths of penetration into the tree structure are visually signalled by vertical bracketed lines (see Fig 2.2 in Chapter 2). To this basic device is then added a minimal range of other visual syntactics which provide the user with a means of indicating programming actions such as iteration, multi-way choice, input and output, escape and parallel actions (the last of these indicating a limited opportunity for breaking away from strictly tree-like structures).

As such tools become increasingly part of software automation it seems likely that the format will play a correspondingly important role in documentation of systems and also therefore in communications between systems analysts and programmers (and indeed the programs themselves where these are the generated outputs of fourth generation languages and application generator tools).

4.5.4 AND-OR trees

Both AND and OR are binary connectives in the sense that they operate on two variables. (NOT is a unary connective since it operates on a single variable.) This can be represented visually by giving the connective two arms as shown in Fig 4.18(i). Since the aim is to model expressions, we require the position at the head of the arms to be reserved for the expression label and therefore we need some visual device to distinguish one connective type from another. One convention is the use of an arc between the arms to indicate AND, the absence of the arc being taken to imply OR (see Fig 4.18 (ii)).

We can now use the technique to draw the structure of a more complex expression (see Fig 4.19). If we use algebra (or some other device) to normalise this expression we get

$$E = ((a \,\&\, b) \vee (c \,\&\, d)) \,\&\, (a \vee b)$$
$$= (a \,\&\, b) \vee (a \,\&\, c \,\&\, d) \vee (a \,\&\, b) \vee (b \,\&\, c \,\&\, d)$$
$$= (a \,\&\, b) \vee (a \,\&\, c \,\&\, d) \vee (b \,\&\, c \,\&\, d)$$

for the disjunctive normal form.

It is possible to read off this normal form directly from the AND-OR tree by following a procedure of spotting alternative minimal sub-trees of the main tree which are sufficient to make the root expression TRUE. We can drop OR branches, but not AND branches. Thus the first such sub-tree is

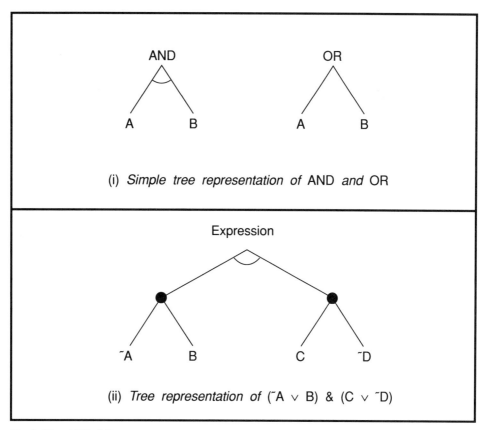

(i) *Simple tree representation of* AND *and* OR

(ii) *Tree representation of* (˜A ∨ B) & (C ∨ ˜D)

Fig 4.18 AND-OR tree representation

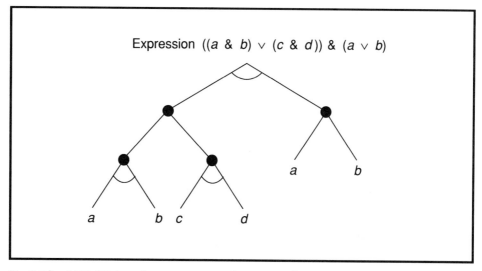

Expression ((*a* & *b*) ∨ (*c* & *d*)) & (*a* ∨ *b*)

Fig 4.19 AND-OR tree for a more complex expression

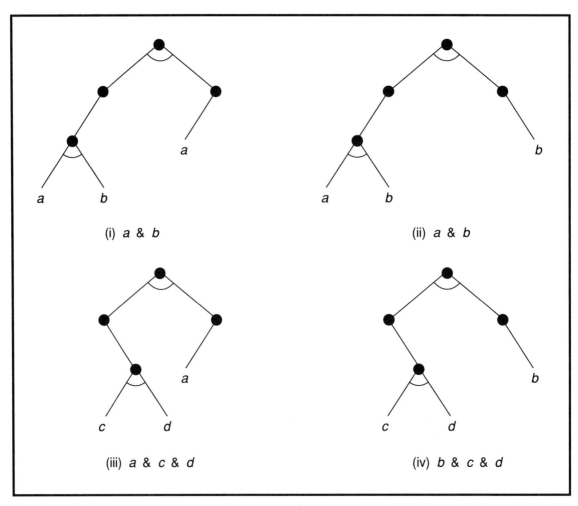

(i) *a* & *b*

(ii) *a* & *b*

(iii) *a* & *c* & *d*

(iv) *b* & *c* & *d*

Fig 4.20   The minimal sub-trees of the tree in Fig 4.19

shown in Fig 4.20(i). Since the branch leading to *a* alone is obviously redundant, we can read this sub-tree as providing the clause {*a* & *b*}. The sub-tree in Fig 4.20(ii) only provides the same clause. The sub-trees in Figs 4.20(iii) and (iv) provide the clauses {*a* & *c* & *d*} and {*b* & *c* & *d*} respectively. These are the clauses we obtained above by algebraic means.

We shall see later that the normal form is an important basis for problem-solving in logic-based knowledge processing and it is sometimes easier to understand the search for a solution in terms of visualising the sub-trees of the AND-OR tree.

4.6 Networks, graphs and matrices

4.6.1 Introduction

Had ours been a more rigorously formal and mathematical approach to maps and charts, this chapter should have started with this section (and a much more detailed version at that) rather than concluding with it. In dealing with charts and tables, what we have been about, taking a more abstract view of things, is the 'making of connections'. Fundamentally, the modelling of a connection comprises, first, at least two things to connect and, second, the connection itself. Put together any number of these connections and one is certain to arrive at a network-like result.

Now mathematicians have long since noticed the fundamental importance of the **network** and have developed a *theory of nets* accordingly, which aspires to make networks manipulable by representing them as symbol strings which then have certain rules of operation applied to them. Rather confusingly for the layman, the theory is also known as *graph theory* (very little to do with graphs as either businessmen or scientists know them).

4.6.2 From network to graph

Let us start with a diagram which is easy enough to recognise as a network of interconnections as in Fig 4.21. As a real and useful diagram this may have come from a representatioin of a whole range of actual situations in the real world. Here are just a few possibilities from the system analyst's environment alone.

A computer network
A document flowchart
A data flow diagram
A program logic chart
Message-exchanging channels between real-time programs
A menu structure for a software interface
A project control chart (PERT, PNA, etc).

We might stop there and say there is no need to go any further, since in each application the actual diagram would be annotated in ways that meet the special needs of that application, and if any processing of the diagrams is required then there will be special software to do it for us.

But we shall not stop there. The fact that each application does share a recognisable diagrammatic form with all the other applications cries out for something to be said about that common form. If indeed there are manipulation techniques which enable certain results to be established quickly and efficiently when applied to the form as a generalised network

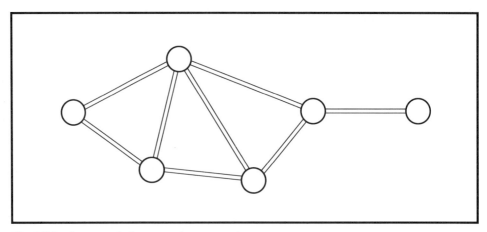

Fig 4.21 A general diagram of a network

—and there are—then should we not at least have some understanding of that, so that we might gain insight into the uses we make of such networks within the specialist applications?

First of all, we should have some way of describing the general network form so that we can develop a relatively formal and consistent model of it. Mathematicians have done this and they have called the consequent system **graph theory** or sometimes **net theory**, as we have already said. (Unfortunately, we again come across the aggravation of alternative terminologies.)

While, in an applied network, the things to be connected and the connections themselves may be thought of as having some substance and might therefore be shown as boxes connected by pipelines as in Fig 4.21, graph theory abstracts these to points connected by lines since it is these that provide the essence of the network as a network. The graph for the network in Fig 4.21 is therefore as shown in Fig 4.22.

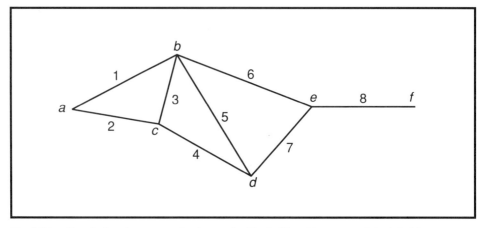

Fig 4.22 Graph for the network shown in Fig 4.21 with appropriate labelling

This is described as comprising **nodes** or *vertices* connected by **arcs** or *edges*. (We shall use the first of the two alternative terms in each case.) Two nodes which are connected through an arc are said to be *adjacent*, and two or more arcs which meet at a common node are said to be *incident* (an adjective, note). The total number of arcs incident at a given node is called the *degree* of that node. Thus in Fig 4.2, node *f* has degree 1; node *a* has degree 2; *c, d* and *e* have degree 3; and node *b* has degree 4.

Adjacency and incidence are of fundamental importance in the setting up of manipulatable models of graphs, as we shall see in the next section. But even at this pre-modelling stage we begin to be able to develop rules about graphs. Here are three which are easy to check (we do not prove them):

- The sum of all the node-degrees is always even and equal to twice the number of arcs. (In Fig 4.22, this is $16 = 2 \times 8$.)
- There are always an even number of nodes of odd degree. (In Fig 4.22 this is 4.)
- In a graph having every node of equal degree (called a *regular graph*), twice the number of arcs is equal to the product of that degree with the number of nodes. (See Fig 4.23, where this is $2 \times 15 = 3 \times 10$.)

Although these are simple enough rules (though tricky to prove) they illustrate the potential for such rules as labour saving devices when carrying out computations based on handling networks. In a large network, for

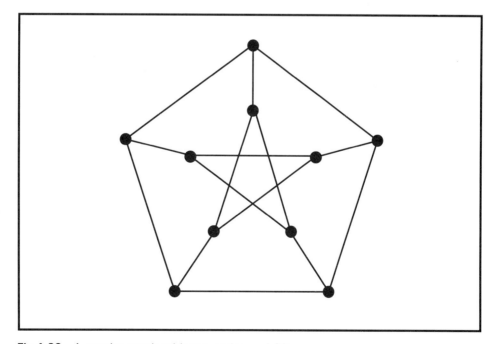

Fig 4.23 A regular graph with ten nodes and fifteen arcs

113

example, it might be a good deal easier to compute the total number of arc-to-node connections by counting the arcs (and doubling) than by counting those connections themselves.

4.6.3 From graph to matrix

A **matrix** is a formal way of setting out values where the position of each value in this **array of values** is of significance. What that significance is will depend upon the application, but the advantage in achieving a matrix arrangement is that an algebra of immense power exists which enables the matrices to be processed mathematically. Matrix algebra is beyond the scope of this text but we can demonstrate easily enough how the matrix form of representation can be used to model graphs.

Two simple approaches involve using the concepts of adjacency and incidence respectively. For an **adjacency matrix** to be formed we notionally label the columns and rows with the node identities and indicate a connection with a 1 and a lack of connection with a 0. The matrix thus formed for Fig 4.22 is shown in Table 4.8(i). The resulting matrix is automatically what we call a *square matrix* since it must have an equal number of rows and columns.

The **incidence matrix** is formed by assigning arc identities as the column labels and node identities to the row labels and indicating that an arc meets a node by inserting 1 in the appropriate position as before. The matrix thus formed in Fig 4.22 is shown in Table 4.8(ii). This is in general a *rectangular matrix* since the number of arcs and nodes is not usually the same.

Notice that the degree of each node is given by the sum along each row. In the adjacency matrix, Table 4.8(i), there is a mirror-imaging around the diagonal matrix and this helps us to see how the rule that the sum of the degrees is twice the number of arcs comes about.

Networks where the direction of flow associated with the connections is significant are very common in applications and it is worth noting that this is easily accommodated by the matrix notation. Suppose there are flow directions associated with Fig 4.22 which give rise to the directional graph,

Table 4.8 *The adjacency (i) and incidence (ii) matrices for Fig 4.24*

	a	b	c	d	e	f		1	2	3	4	5	6	7	8
a	0	1	1	0	0	0	a	1	1	0	0	0	0	0	0
b	1	0	1	1	1	0	b	1	0	1	0	1	1	0	0
c	1	1	0	1	0	0	c	0	1	1	1	0	0	0	0
d	0	1	1	0	1	0	d	0	0	0	1	1	0	1	0
e	0	1	0	1	0	1	e	0	0	0	0	0	1	1	1
f	0	0	0	0	1	0	f	0	0	0	0	0	0	0	1

(i)   (ii)

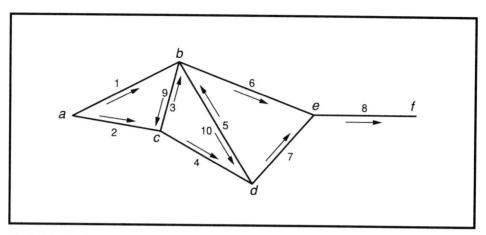

Fig 4.24 A digraph version of the graph in Fig 4.22

or **digraph**, shown in Fig 4.24. For the adjacency matrix we merely adopt a convention which says that columns indicate 'to' and rows 'from' (indicated by the arrow in Table 4.9(i)). The redundancy of the mirror images around the diagonal is now used up by having each triangular half of the matrix devoted to each of the two directions.

A little more ingenuity is required to devise a way of having the incidence matrix covering the digraph, since there isn't the same kind of redundancy in the modelling. One way is to indicate 'from' as positive and 'to' as negative. Table 4.9(ii) should be read column by column to check that it is in this way a true representation of Fig 4.24.

Readers familiar with, say, PERT, techniques will be aware of concepts such as shortest paths through a directed network. In PERT the 'length' referred to is to do with the fact that the arcs have time-period values attached to them. Although many such methods depend upon rules of thumb, the discipline provided by the matrix representation can be invaluable in automating the appropriate algorithms.

Table 4.9 *The digraph adjacency (i) and incidence (ii) matrices for Fig 4.24*

→a	b	c	d	e	f
a	0	1	1	0	0	0
b	0	0	1	1	1	0
c	0	1	0	1	0	0
d	0	1	0	0	1	0
e	0	0	0	0	0	1
f	0	0	0	0	0	0

(i)

	1	2	3	4	5	6	7	8	9	10
a	1	1	0	0	0	0	0	0	0	0
b	−1	0	−1	0	1	1	0	0	1	−1
c	0	−1	1	1	0	0	0	0	−1	0
d	0	0	0	−1	−1	0	1	0	0	1
e	0	0	0	0	0	−1	−1	1	0	0
f	0	0	0	0	0	0	0	−1	0	0

(ii)

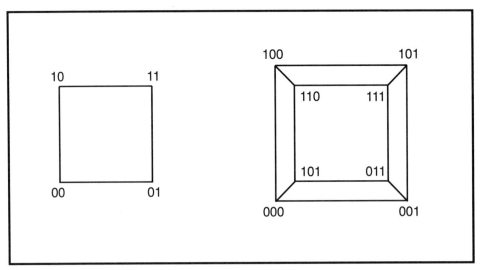

Fig 4.25 The k-cubes for two- and three-bit codes

4.6.4 Matrices in logic

Although this takes us into quite difficult-looking areas of work it is worth having some idea of the scope for the use of matrices in logic work. The first we mention is the graph system that can be generated by representing binary codes as graphs. Fig 4.25 shows the first three systems associated with two-bit and three-bit codes respectively. The graphs can be read as having nodes placed at the centres of the k-map cubes—the graphs are called k-cubes—so moving from one node to the next always involves a change in only one bit.

This has all sorts of possibilities in using graph methods supported by matrix representation to explore the behaviour of codes.

But perhaps the most promising development is the creation of a *matrix logic* which sets out to interpret the k-map definitions of the Boolean connectives as mathematical matrix objects. The connectives then become matrix operators with great operational power in matrix algebras of wider scope. The discipline is still in its infancy and is an area which should be the attention of much fundamental research in the next decade or so.

5 Sets, Lists and Relations

5.1 The meaning of sets

5.1.1 Introduction

Like many of the most elegant areas of mathematics the theory of sets is a striking combination of simplicity and power. The simplicity is the basic idea of what is meant by a set and the power is in what this view enables us to do with sets using mathematical method and, more importantly to us here, automatic programming. The tackling of problems which are most suited to the application of set theory without using that theory is rather like trying to run a mailing service without the use of labelled envelopes.

And that is a useful metaphor from which to launch our introduction because, as much as anything, set theory first of all is to do with appropriate labelling. Quite simply, we can find we have a set to work with because we have chosen to label something as a set. As its name suggests, the label is instinctively applied to a collection of things. These things can be objects, events, ideas, hieroglyphs or whatever you wish, but as long as you can clearly define what is part of your collection then you have a set. Of course 'clear definition' is contentious because there is a degree of subjectivity in that term, but the point is that such ideas have to be related to context. What passes as clear enough in one context may not be clear in another.

Also, the idea of a set as a kind of physical collection of objects is misleading in two respects. One is that while it is possible to have a physical set which contains clones, such as the set of red balls in a snooker set, the set concept does not permit duplicate names for elements within it; each red ball would have to be given a different identity or name in order to form such a set. Once something has been named as being in a set (red ball) then it cannot be designated a second time in that same set.

The second difference is that the sequence of naming or listing the elements is immaterial. Once elements are in a set the bag can be shaken up as often as you like; as long as no members are lost and none gained it remains the same set. This is only true of a physical set if one deliberately ignores physical position.

So how can we 'define' a set? There is a variety of ways to choose from.

Here are the more common ways:

a By identifying members explicitly. In a physical sense this is done by pointing to and naming each in turn, eg. "My set comprises that box, the digital number 14, the month of January and the name Alwyn, and I shall call the set My Birthday!"

This set comprises a physical object (the box), a widely understood symbol (14), a fairly abstract concept of time (January) and a character string (Alwyn). Quite a mixed bunch it may appear, until the chosen name is declared and the set then makes more coherent sense. It didn't strictly have to make such sense, just so long as the definition was clear enough to enable the observer to recognise the set and name it in the same way.

b By citing a characteristic feature or features of the whole membership which acts as an unambiguous definition of that membership, eg. "My set is the collection of all living four-legged mammals and I shall call it 4-legs."

This introduces an interesting problem, namely that although the definition may be accepted as clear enough it is pretty well guaranteed that no-one would be able to point at every member of the set! This does not invalidate its candidature however. It may have to remain as an idea for a set, but it is still a set and can be processed as such by the mathematics. Some apparently simple set operations, though, will be impossible to resolve in reality, eg. "Count the number of members in the 4-legs set." Notice the resemblance to issues of language here: "I can talk about it, but don't ask me to do it!".

c By appealing to labelling conventions already in use, eg. "My set is all the character-strings which represent the full names of the months of the year, and I shall call it Month-names."

This definition appeals to a common understanding of the phrase 'full names of the months of the year'. 'Full' is desirable in case it is thought that abbreviations would do and, since there are may versions of such abbreviations, we may find that the definition has failed to pass the unambiguity test. It is also of interest because the set is a much more cohesive collection of data than before. Information technologists can recognise it as a storable table or list, and may even begin to ask operational questions about it such as: "Do you want it stored in alphabetical sequence or in temporal sequence?" But remember that at this stage it doesn't matter about the sequence; it is sufficient that we have managed to define the set membership clearly enough.

d By calling up a definition from an existing system of thinking, the most common perhaps being that of the field of mathematics, eg. "My set is all the positive integers and I shall call it Posint."

What this definition introduces is the idea of a set of truly infinite membership. Its members are not merely physically impossible to count, they are understood by mathematicians to be literally countless. And yet, again, this is an acceptable set and amenable to many of the mathematical set operations that are used with non-infinite sets.

Having identified a set satisfactorily, there is still the problem of being clear about how to write down its description. We shall consider more formal notation in a following section, but here let us consider again the importance of labelling. Our sets are usually written down for human communication purposes (whatever the method of storing them, eg. within computer systems) and it is important to be clear about whether the set member is meant to be the object itself, its accepted identification or name, or the string of characters representing that name. The confusion arises because, whichever of these is the case, we still use symbol strings in the final event to write down the identification. The example under **a** above was intended deliberately to raise awareness of this issue.

Apropos of this, the reader might think for a moment about the well-known children's conundrum about finding a sentence containing a consecutive sequence of seven 'and's. The solution concerns the sign being painted on the shopfront of the brothers whose family name is And. Looking at the sign one brother says to the other "Wouldn't you agree that the spaces between And and & and & and And are unequal?" and his brother replies "You're right." (The 'reply' coda has been added to provide one more type of 'and' label.) Actually, the sequence can be lengthened by introducing more brothers but the above gives us sufficient variety in types of 'and' labelling to make the point.

5.1.2 Special sets

There are two sets which are as essential to set theory as the numbers zero and unity are to number theory. The *empty set* or **null set** is the set which comprises no members. One important way in which it is special is that it is potentially a part of each and every set-related problem. The empty set in, say, a mathematically oriented problem is no different to the empty set in a problem related to, say, collections of objects in a warehouse.

The other special set is the **Universal set** which is that set which contains as members every member of every set in the problem area under discussion. This is special in that it is a kind of problem boundary-marker, as its rather grand full name of 'universal set of discourse' indicates.

A set may of course comprise only one member—sometimes called a *singleton set*—which has the important consequence that each and every member of a set can be regarded as a set in its own right. Any set is therefore inevitably a set of sets. In any case, each element might also be a label for another set, in which case the parent set is most certainly a set of sets.

Another very important concept is that of the **subset**. This is fairly generally understood, without recourse to formal definition, as any set comprising only some of the members taken from the set of which it is a subset. But this disguises the more formal truth that any set is a subset of itself! Also, more understandably, the null set is a subset of every conceivable set, including itself. So now we see that any set can also be used as the basis for generating all its subsets and this resultant set of sets collection is also a rather special set—often called the *power set* of the set from which the collection is generated.

Consider the set comprising 'you' and 'me'. The power set of this set is the set comprising the following:

'you' and 'me'
'you'
'me'
'nobody' (the null set).

5.1.3 A symbolic language for set theory

As the previous section was beginning to demonstrate, natural language can quickly become a confusing way of working with sets (just as it is for working with numerical values). We shall therefore introduce some basic conventions for working in sets in a succinct and efficient manner. For the non-mathematically inclined this can quickly begin to look awesome, but at this level we really are just introducing a kind of shorthand—and it is not as difficult for the reader to translate back into natural language as it may at first seem. This should in any case be a familiar enough process to anyone who has done any kind of programming.

Conventions for writing out a set do vary, unfortunately. However, some general pointers will help to overcome any variation encountered. What we have essentially is a list of *elements* or *members* (it is convenient to retain these two alternative nomenclatures) which have to be embraced visually in some way. Diagrammatically this can be shown by drawing a boundary line around the named elements (see Fig 5.1), a convention which we have already used when introducing Venn diagrams with Boolean algebra in Chapter 2. However, this is not very convenient for written work and the symbol manipulation of mathematics, and therefore the convention of enclosing a horizontal list within brackets is adopted. Conventions vary as to whether the elements are separated by spaces or commas and as to what type of brackets are used. We shall adopt comma-separation and curly braces { }.

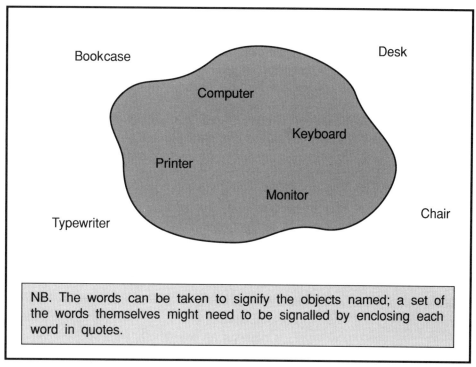

NB. The words can be taken to signify the objects named; a set of the words themselves might need to be signalled by enclosing each word in quotes.

Fig 5.1 Depicting a set by mean of a boundary line

For example, the set called My Birthday! in paragraph **a** of Section 5.1.1 would by our conventions be written:

{Box,14,January,"Alwyn"}

We have already been using labels for the set as a whole, but it is advisable to adopt conventions regarding that naming, especially where computer usage might be implied. Indeed the convention may be dictated by the programming environment, and these do vary considerably. **Bold type**, *italics*, UPPER-CASE lettering are all possible binding conventions in different environments. We shall adopt the convention that a set name must start with an upper-case letter and not contain spaces; thus for the single-symbol labelling preferred by mathematicians, a single capital will imply that it is the name of a set (so long as the context implies it too of course). This modifies our usage of lower-case in Chapter 2, where direct comparison with the Boolean variables was the order of the day. In the set theory context we shall reserve lower-case for representing the elements within a set, a useful visual contrast.

The null set is also represented by a number of different conventions, the most common being: zero 0, zero in bold type **0**, zero with a slash through it Ø, and empty brackets. We shall adopt the last of these so that { } will nicely depict the empty set, while not forgetting that zero may still be useful when looking at the Boolean and general logic context.

The universal set of discourse, or more briefly Universal set is less important to nail down with a single symbol since its contents do vary with context. However, it can be useful to emphasise its role and we can adopt the use of the upper-case U on such occasions. (W for World is also used.) Again, in the appropriate context we can revert to the 1 of Boolean logic.

Two phrases that we shall want to use very often are: *is a member of* and *is a subset of*. There are widely accepted standard symbols for these which are $\in$ and $\subset$ respectively. Note that they do serve very similar purposes in that an elemental member of a set is automatically a subset of it, but often it is essential to emphasise the member role.

Two more phrases that we have not yet had need to use, but which do play very important roles in the discussion of sets, are: *For every element* ... and *There is at least one element* ... The symbols adopted for these, $\forall$ and $\exists$, do seem to non-mathematicians somewhat perversely chosen but they are almost universally the conventions in use and therefore we must stick to them. They are in fact called the **quantifiers** and are mentally read out in the abbreviated forms:

$\forall x.$ For All x such that ...

$\exists x.$ There Exists an x such that ...

The symbols do in fact aid the memory—the inverted A for the All in the *universal quantifier* and the reversed E for Exists in the *existential quantifier*. Because the reference is to element or elements it is the lower-case x that is used rather than the upper-case, which we have agreed should depict whole sets. The full stop after x in each case helps to separate visually the quantification symbols from whatever statements may follow.

As examples of using these symbols, consider the following:

$$\forall x. \ (x \in A) \rightarrow \ \tilde{}(x \in B)$$

This should be read as: For all elements which are members of the set A, it follows that they are not members of the set B. This tells us that A and B share no elements as members—they are in fact what are known as **disjoint sets**.

$$\exists x. \ x \in (A \& B)$$

This should be read as: There exists an element which is a member of both set A and set B. This tells us that the overlap or *intersection* between the two sets is not the empty set.

5.2   The logic of sets

5.2.1   The sets-and-logic connection

In the examples at the close of the previous section we have used the Boolean connectives NOT ($\tilde{}$) and AND (&). In Chapter 2, in our introductory survey of logic forms, we used classes and sets to describe the meaning of Venn diagrams. But here, in dealing with sets, we have introduced operations that did not appear before. What is the significance of this?

First of all it should be fairly clear that in our two-valued Boolean logic discussions we always treated the variables as whole entities, with the characteristic that they could adopt one of two states: T/F, ON/OFF, 1/0, etc. But, although a set is a whole it is also a collection of constituent parts, and it is the relationship of those parts, the elements, with the whole, the set, that provides the two-valued characteristic, namely MEMBER/NOT-A-MEMBER. Now, although membership is indeed a two-valued quality and must therefore be a satisfactory candidate, as it were, for two-valued logic treatment, there are aspects of set membership that are beyond the two-valuedness feature.

Does this set have any elements as members? Is this element a member of this set? How many elements does this set contain? These questions which may be asked of any set are clues to some of these 'other' features. We also know that in sets of records in data files, or for that matter sets of characters in a character-string, the sequence of elements is highly significant. Further questions arise then. What sequence? How to add and delete elements? How to change sequence? These basic file processing issues again do not seem to fit safely within the two-valued logic view of things.

Although in order to cope with sequence we need to extend set theory into the domain of list processing as we shall see, at least set structures are closer to the practical world of data processing than is pure Boolean logic. In this respect they are crucially important to us as a first step in perceiving *the logic of the processing of data*. The steps that we can then follow beyond simple sets take us into lists, relations and predicated forms, which are the next spans of the bridge over into the representation and processing of the more sophisticated knowledge representational forms of data.

Finally, let us return to our straight Boolean variable—the 'whole' which must be in one of two states. When we say

Let *a* be the Boolean variable for ...

we are in fact implying that the symbol stands for a possible host of cases which fit the behaviour pattern of the case we are modelling. For example, if *a* stands for the statement "the program has been tested", until we have pinpointed precisely which program, the statement could potentially stand for all programs! Even if the context is a specific one, where it is intended to construct an instruction about what action should follow, it is likely that *a* could stand for "this program and all programs like it". Another point of view is that the program itself comprises a set of constituent sub-progams. Therefore, as soon as we consider an apparently 'whole' Boolean variable we find that it is possible that we (i) wish to generalise about it as being typical of a set of like cases, or (ii) wish to dissect it to see what it's made of. In the formal logic sense set theory is the key to doing this.

5.2.2 Working with sets

We should be able to go straight into the Boolean type of operations now without much explanation being necessary. Here are brief notes on the main ones.

The AND operation between sets is variously referred to as their *intersection*, meet or overlap. The intersection of two sets results in a set whose elements are those which were members of both the operand sets, and only those elements. Because of the definition of the set form, no duplicates must appear in the resultant set—hence the use of the term 'overlap' for this intersection. Thus, for example:

1 If P = {India, Africa, America} and Q = {Europe, America, China, Africa}, then P & Q = {Africa, America}. Note that sequence is still immaterial at present.

2 If X = {2, 4, 6, 8} and Y = {1, 3, 5, 7}, then P & Q = { }, the empty set. P and Q are disjoint.

3 If A = {IBM, Bull, Apple} and B = {Apple, Bull, IBM}, then A & B = A = B.

Note that the symbols ∩ and ∪ are also used for the AND and OR connectives respectively, especially in texts dealing exclusively with sets.

The OR operation between sets is referred to as their *union* or join. The union of two sets results in a set whose elements are those which are members of either or both of the operand sets. (Remember, this is inclusive-OR.) Thus, using the same example set as above:

4 P ∨ Q = {Africa, America, China, Europe, India}

5 X ∨ Y = {1, 2, 3, 4, 5, 6, 7, 8,}

6 A ∨ B = A = B

The NOT operation on a set requires that the Universal set be known in order that the *complementation* can be carried out, since the resultant set contains all the elements which are not in the operand set. Thus for example:

7 If U = {a, e, i, o, u} and A = {a,e,i}, then ˜A = {o, u}. We would not be able to give the explicit expression of ˜A did we not know that the Universal set in this context was 'all the vowels'.

8 Staying with the vowel-set A above, we can see that

$$A \& \~A = \{ \quad \}$$
$$A \lor \~A = U$$
$$A \& U = A$$
$$A \& \{ \quad \} = \{ \quad \}$$

and so on.

These must of course always be true anyway and because U and { } are the sets equivalent of T and F, 1 and 0, and so on. We could make that equivalence even more explicit by reading U as "True for all cases (elements)" and { } "No cases for which it is True (ie. always False)".

Our sets might be classes of elements, whose membership may or may not be physically countable individually. For example:

9 Let X represent the class 'All female people' and Y represent the class 'All programmers', then:

$\~X$ represents the class of non-female (male?) people
$\~Y$ represents the class of non-programmers
$X \lor Y$ represents female programmers
$X \& \~Y$ represents female non-programmers.

and so on. What the Universal set is would need to be defined of course. According to problem context it could be any one of an infinite number of possibilities, such as

the whole population of the world
the population of the nation
the personnel of a company
an IT Department's personnel

though it must be wide enough to encompass the class variables in some way or the formulation becomes pointless.

So long as we are able to treat sets as wholes or classes without need of reference to individual members, we can then work in an identical way to Boolean logic. For example, continuing with the previous sample class sets:

10 Suppose U is the UK population and a certain company decides to adopt the policy that it will seek, as systems analysis recruits: "male programmers, female programmers, or female non-programmers". This can be expressed algebraically as

$$(\~X \& Y) \lor (X \& Y) \lor (X \& \~Y)$$

which simplifies (by any method one may care to choose) to

$$X \lor Y$$

which means: "females or programmers (or both)". So a clearer statement of the policy might be: "We shall seek to recruit females as systems analysts and also programmers of either sex", a re-formulation perfectly achieveable without the algebra no doubt, but the example helps to demonstrate the point we had made about classes as Boolean variables.

5.2.3 Implication and sets

There are some interesting correspondences between the IMPLIES connective ($\rightarrow$) and certain properties of sets which greatly help to visualise implication diagrammatically. We touched upon this in Chapter 3, Section 3.3.2, when we illustrated implication with a special Venn diagram. The diagram is reproduced as Fig 5.2, but the circles should now be read as representing sets and labelled here X and Y.

We can see that this also illustrates the subset property $X \subset Y$. It is common sense that if an element is a member of X then it must also be a member of Y when $X \subset Y$. In other words,

Membership of X IMPLIES Membership of Y

The algebra shows the correspondence too, for in Fig 5.2 we can observe that the set resulting from $\{^\sim X \vee Y\}$ is the Universal set, since the $^\sim X$ brings all of $^\sim Y$ into the result and the $\vee Y$ joins all of X back into the result! So when $X \subset Y$ then $\{^\sim X \vee Y\}$ is true for all cases. (We could also have observed that $\{X \,\&\, ^\sim Y\}$ is the empty set, and worked from there.) Indeed, as long as we have been careful to define symbol conventions, we can write down the equalities:

$$X \rightarrow Y = X \subset Y = {^\sim}X \vee Y = {^\sim}(X \,\&\, Y)$$

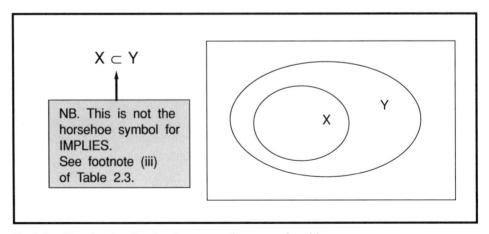

$X \subset Y$

NB. This is not the horsehoe symbol for IMPLIES.
See footnote (iii) of Table 2.3.

Y
X

Fig 5.2 Showing implication in terms of set membership

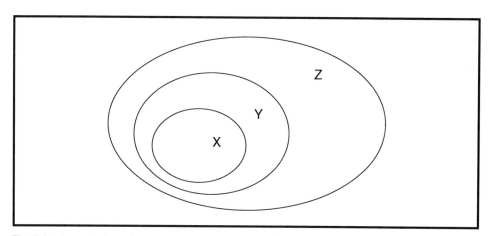

Fig 5.3 The chain rule illustrated by set membership

Apart from its intrinsic interest, the correspondence is, as mentioned before, very useful for sketching pictures of implication expressions. For instance, the *chain rule*, which we shall look at more closely in Chapter 8, is immediately visible in the sketch shown in Fig 5.3.

If $X \to Y$ and $Y \to Z$, then $X \to Z$

In Chapters 2 and 3 we gave stern warnings about the confusions occurring between implication and equivalence. With set illustration, the difference is much easier to see. Equivalence is, in fact, the special case where a subset of a set is the set itself! If $X \subset Y$ and $Y \subset X$, the only explanation is that they are identical sets and that membership of each implies membership of the other. We wrote that as $X \leftrightarrow Z$ in Chapter 3. The picture here helps to emphasise that $X \leftrightarrow Y$ is just the same as $X = Y$.

But there is more. The disjoint condition that we have already met, ie. where two sets share NO common elements, can be illustrated as shown in Fig 5.4.

We can see—quite literally see—that X is a subset of $\tilde{}Y$ and can immediately conclude therefore that $X \to \tilde{}Y$. But it is also true that Y is a subset of $\tilde{}X$ and therefore $Y \to \tilde{}X$. Since they come from identical formulations, it must be that

$X \to \tilde{}Y = Y \to \tilde{}X$

This is a perfectly valid transformation—a kind of De Morgan Rule of implication. Put in slightly different form:

$X \to Z = \tilde{}Z \to \tilde{}X$

(This can be obtained either by replacing Y with $\tilde{}Z$ or by direct observation of Fig 5.2.) It is a very important transformation in knowledge processing as we shall see later, for it tells us that if we prove a false X from a false Z it is just as valid as proving a true Z from a true X! In effect it gives a

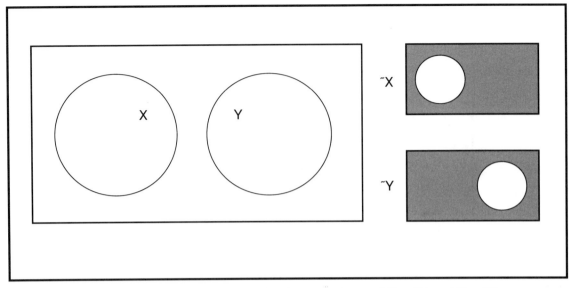

Fig 5.4   Disjoint sets illustrating the equivalence of $X \to {}^{\sim}Y$ and $Y \to {}^{\sim}X$

knowledge processor an alternative line of attack in seeking to find valid proofs.

5.2.4   Differences and products

Figures 5.2 and 5.4 show special conditions obtaining between two sets. What if the conditions are 'normal' and there is some overlap between the two sets, as a general Venn diagram would illustrate? Figure 5.5(i) shows such a situation.

We shall stipulate here that the overlap is not the empty set, or

$$\exists w. \; w \in X \& Y$$

There is a DIFFERENCE connective (a minus sign can be used) that can be

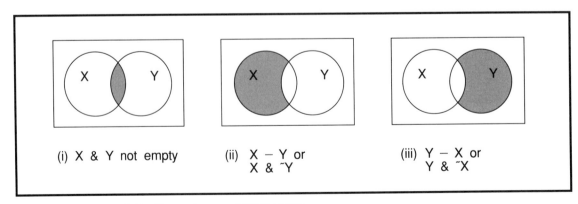

(i) X & Y not empty

(ii) X − Y or
X & ${}^{\sim}$Y

(iii) Y − X or
Y & ${}^{\sim}$X

Fig 5.5   Illustrating set DIFFERENCE

defined between two sets. Fig 5.5(ii) shows the set resulting from X – Y and Fig 5.5(iii) shows Y – X. The 'subtraction' is quite literal—elements which are common to the two sets are removed from the set which is being subjected to the subtraction, ie. the set preceding the DIFFERENCE sign.

But these resultant sets have algebraic intrepretations too, which are easily obtained, from the K-maps for example, and are

$$X - Y = X \,\&\, {}^{\sim}Y \quad \text{and} \quad Y - X = Y \,\&\, {}^{\sim}X$$

By observing the subset relationships in the three diagrams of 5.5 we can see that

X & Y is a subset of both X and of Y
X – Y is a subset of both X and of ˜Y
Y – X is a subset of both Y and of ˜X

There are therefore many implications that are always true for a pair of sets (remember these sets are not special, being neither subsets or disjoint). These are

X & Y → X
X & Y → Y
X & ˜Y → X
X & ˜Y → ˜Y
Y & ˜X → Y
Y & ˜X → ˜X

But, in truth, all these are simply variations on the first two in the list. If a conjunction between two logic variables is true, then it follows that each of the variables is separately true. This is easily accepted as common sense, but again the use of set-sketching can help to guide one through the understanding of more complex combinations.

In passing we might note that the DIFFERENCE notation produces an interesting form for the constant rules. For instance, it should be obvious that

$$U - X = {}^{\sim}X$$

This is of course just $1 \,\&\, a = a$ in disguise!

While DIFFERENCE is easily mentally associated with the concept of arithmetic subtraction, the PRODUCT of sets seems, in spite of its name, more remote from arithmetic multiplication. The PRODUCT of two sets is defined as an operation which generates a set of sets, the set elements being formed by taking all possible pairings of the elements of the candidate sets. PRODUCT is symbolised in various ways—we shall use the large cross X.

An example of the operation of PRODUCT is as follows:

{Analysis, Design, Implementation, Maintenance}
    X {Time, Staff, Budget}
= {{Analysis,Time}, {Analysis,Staff}, {Analysis,Budget},
    {Design,Time}, {Design,Staff}, {Design,Budget},
    {Implementation,Time}, {Implementation,Staff},
    {Implementation,Budget}, {Maintenance,Time},
    {Maintenance,Staff}, {Maintenance,Budget} }

This example is meant to hint at the way in which PRODUCT might be a useful operation in practice. However, we shall see later that, when the candidate sets are themselves sets of sets and/or lists, PRODUCT can be carried out in different ways and with different results, and a more careful definition of the different types of PRODUCT is then necessary.

It is possible to extend the PRODUCT to cover three (or more) sets, eg. A X B X C, in which case the resulting set is a set of sets in which the element sets contain element triples (and quadruples for four sets, etc.).

A basic simple rule of combinatorial arithmetic is the Multiplication Rule which tells us that the number of different ways of choosing a combination made up of an element from a set of m objects together with an element from a set of n objects is $m \times n$. So now we can see why our rule is called the PRODUCT rule. But note that, should the same element appear in more than one set, the resultant set will contain sets with less than these numbers of elements because duplicates will be dropped.

5.3 The quantification and enumeration of sets

5.3.1 Using the quantifiers

We have already met the universal and existential quantifiers ∀ and ∃. They do not deal with the number of elements as such, but they do 'quantify' in the sense of enabling us to deal in logical terms with the natural language quantifying phrases 'all of the' and 'at least one of the'.

The universal quantifier symbol is often in some respects redundant, because if we are making a statement about a set, then it can be taken as read that we mean the statement to be true 'for all' of the member elements. However, in setting out logical arguments unambiguously it is often desirable to be explicit, especially when we come to *predicate* our statements as will be discussed in Chapter 8.

The existential quantifier, as we have already seen, does need to be used for completeness of expression, such as in

$\exists x. \ x \in (A \& B)$

when wanting to make explicit that the intersectioin of A and B is not the empty set. This example tells us that there exists at least one element x such that x is a member of the set formed by the intersection of sets A and B. If A and B were two files and x any record which may be on either of these files, the statement is asserting that there is at least one record common to both files.

But $\forall$ and $\exists$ can also be important as a pair working together. They are bound together by logical rules which we have not yet made explicit. To demonstrate these rules let us refer to elements e of some universe in which there are sets formed by specifying some feature F. Thus e might be the individuals in an Information Centre and F might be the feature "on the Power Station project." Suppose that some but not all individuals are currently on the project, then we can say:

$\sim(\forall e.\ e \in F)$ which is stating that "it is false to say that all IC staff are on the PS project."

But we could also express the situation by saying:

$\exists e.\ e \in \sim F$ which is stating that "there exists at least one individual who is not on the PS project."

Though we have only demonstrated rather than proved that these are equivalent in this example, the equivalence is indeed always true. Thus,

$\sim(\forall e.\ e \in F) = \exists e.\ e \in \sim F$

If no-one was on the PS project, the following would apply:

$\sim(\exists e.\ e \in F) = \forall e.\ e \in \sim F$

which means that "There is not even at least one individual on the project" is equivalent to "Every individual is not on the project." If everyone is on the project, we would simply have to change the negation of F in this last equivalent-pair:

$\forall e.\ e \in F = \sim(\exists e.\ e \in \sim F)$

We might well react to all this in a natural language context by saying that it is all merely a play on words. But when we find transformations like this in mathematical forms, we know that we are extending our ability to use automation for the logical processing of knowledge. Knowledge processors and especially expert systems will almost certainly make the mathematics more transparent than we have here, but the results are of fundamental importance to the basic appreciation of logical processes.

As a final demonstration, we can show how the above rules work when F is replaced by a logical expression. Let us suppose there are two projects F and G and that there are some individuals who are on both. We then have

$$\exists e.\ e \in (F \& G) = \sim(\forall e.\ e \in \sim(F \& G))$$
$$= \sim(\forall e.\ e \in (\sim F \vee \sim G)) \quad \text{by De Morgan}$$

The final form of this tells us that (excuse the inelegant English for the moment):

> "It is false to say that every individual is not on one or other (or neither) of the projects."

Although it is possible to get there by natural language, this does give a flavour of how more complex statements of this kind can be translated more reliably using the automation of the logic.

5.3.2 Enumeration of sets

If we do want to get down to the actual business of counting elements within sets, this might seem to be the province of pure arithmetic rather than of logic. But there is an area where logic and arithmetic meet and the rule associated with it has important application in the combination of probabilities.

First of all let us introduce some way of symbolising the counting of elements within a set. We shall use the following:

$N(X)$ is the number of elements in the set X.

Thus if $X = \{$Cobol, C^{++}, Fortran, PL1$\}$, then $N(X) = 4$. And if $Y = \{$Prolog, Smalltalk, Lisp, C^{++}, Logo$\}$, then $N(Y) = 5$.

Visual inspection of these two sets of programming language names reveals that

$N(X \& Y) = 1$, since C^{++} is the only common element.
$N(X \vee Y) = 8$, since C^{++} will not occur twice in the result.

These enumerations are related in this way:

$$N(X \vee Y) = N(X) + N(Y) - N(X \& Y)$$

Here, specifically,

$$8 = 4 + 5 - 1$$

The equivalent form of the equality for three sets is

$$N(X \vee Y \vee Z)$$
$$= N(X) + N(Y) + N(Z) - N(X \& Y) - N(Y \& Z) - N(X \& Z) + N(X \& Y \& Z)$$

The counts of the intersections of all possible set pairings are subtracted, while that of the three-set intersection, $N(X \& Y \& Z)$, is subtracted.

In the two-set case we had to subtract the set-pair count, $N(X \& Y)$, because the common elements are otherwise counted twice. Similarly in the three-set case, except that now we have over-compensated by an amount equal to the term $N(X \& Y \& Z)$, which therefore has to be added back in!

The rule is known as the *Principle of Inclusion–Exclusion* and can be generalised for any number of set combinations. Terms involving intersections of even numbers of sets are always subtracted, while those involving odd numbers of sets are always added (including the single-set counts $N(X)$, etc.).

Let us watch the application of this principle in a simple (though complicated enough) three-set example. In Fig 5.6 there are displayed some statistics about a department's employees.

Do the statistics cover all staff? We can make a first check by using the principle of inclusion–exclusion on all the figures after the first row. The first block of three figures are for the single-set counts and the second block of three are for two-set combinations. The final line is for the three-set combination. Therefore,

$$N(\text{All sets}) = 156 + 94 + 132 - 12 - 45 - 24 + 7$$
$$= 382 - 81 + 7$$
$$= 308$$

But this is greater than the total given of 300 staff! There is clearly then some inconsistency in these statistics. Perhaps the total is out-of-date, having not been updated 'at 31 Dec'. Perhaps there is simply a typing error. Had the discrepancy been the other way, we might have assumed that there were 8 people in the department who didn't 'fit the statistics' in some way. But what way? Surely everybody has to be over 24 or not, a UK graduate or not, married or not? The figures would still have been fishy.

Set enumeration can be a useful model for understanding probability computations. The proportion of the number of elements in a set to the number in the Universal set can be interpreted as the probability associated with that set. Imagine a target (the Universal set) with a scaled drawing of a set on it—a set of 20 elements on a target of 100 elements would cover

ANALYSIS OF STAFF BY AGE, QUALIFICATION AND MARITAL STATUS	
Total number of staff	300
Staff over 24 at 31 Dec	156
Number of UK graduates at 31 Dec	94
Number of married staff at 31 Dec	132
UK graduates over 24 at 31 Dec	12
Married staff over 24 at 31 Dec	45
Married UK graduates at 31 Dec	24
Married UK graduates over 24 at 31 Dec	7

Fig 5.6 Data for worked example in set enumeration

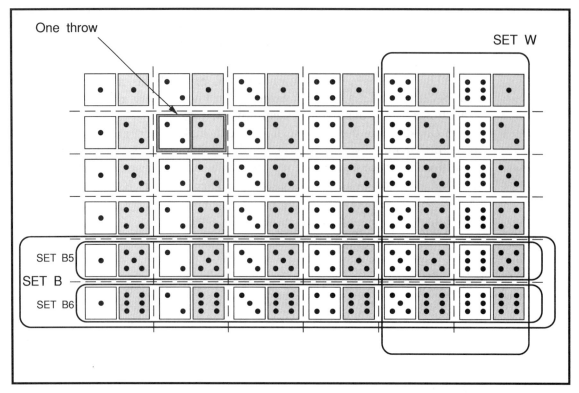

Fig 5.7 All possible throws of a pair of dice

one-fifth of the target area—and imagine randomly distributed missiles striking the target. Each missile has a one-in-five chance (0.2 probability) of hitting the set, so the set can be taken to represent an event with that probability of occurrence. A second set, if disjoint from the first, would represent the probability of an event which can only occur separately from the first event. If not disjoint, the intersection set would represent the probability of both events occurring together.

To demonstrate this in a very literal way look at Fig 5.7. It shows all the possible outcomes of the throw of a pair of dice, one black and one white. The probability of throwing a five on black, set B5, is $N(\text{B5})$ ($=6$) in 36 and of throwing a six on black, set B6, is $N(\text{B6})$ ($=6$) in 36. These two events cannot occur together (disjoint sets), so the probability of a black five or six, the set B, is the sum of the separate probabilities, ie. 12 in 36.

Similarly, the probability of throwing a five or six on white, set W, is $N(\text{W})$ ($=12$) in 36. But the probability of any five or six appearing in a throw is

$$N(\text{B} \vee \text{W}) = N(\text{B}) + N(\text{W}) - N(\text{B} \& \text{W})$$

ie. $12 + 12 - 4 = 20$ in 36, since the same numbers can appear simultaneously on both dice (non-disjoint sets).

134

This example is deceptively neat and tidy. Dice are assumed to be perfectly cuboid and of uniform density (not loaded). The black and white dice are identical. Each number on the dice faces appears only once on each die. Throws are always executed perfectly (never an off-the-table throw where the dice becomes lost or stuck awkwardly). Life is never as simple as that, but instructive illustrations often need to be simple to convey their messages. We shall leave until later the issue of the need for knowledge modelling to be able to cope with life's less-tidy features.

5.4 Lists

5.4.1 Ordered sets

We have stipulated that the sequence of elements in a set is not significant. We can however tighten up the definition to incorporate significance of sequence. Indeed, in view of the fact that sequence and arrangements of data are of great importance in the automation of data handling, it is important that theory should be prepared for this constraint.

If we were to extend every element in a set to comprise two pieces of information—its own content and an ordinal value indicating its sequential position among its fellows—then it will be clear that two sets such as {Red, Amber, Green} and {Green, Amber, Red} are no longer equal sets, because each element has an implicit differentiator associated with it: Red-1 is not the same element as Red-3, etc. In fact, since we have in general to write down (or store) elements in a left-to-right sequence anyway, we can easily accept such sets as not being equal merely by visual inspection (or by program scan).

It is useful to provide some visual distinction between normal (unordered) sets and ordered sets especially when they appear within the same text or system. Let us adopt the convention for ordered sets of placing spaces (not commas) between elements and round (not curly) brackets.

So, by this convention, we would have

{Red, Amber, Green} is equal to {Green, Amber, Red}

since these are normal sets. But

(Red Amber Green) is not equal to (Green Amber Red)

The most basic ordered set is the ordered-pair, encountered in any co-ordinate system of reference. The $(x\ y)$ co-ordinates of two-dimensional graphs are familiar enough to all students of school algebra. They are often known as Cartesian Coordinates after the great mathematician and philosopher Descartes. It is sets of these ordered-pairs, defined by a constraining functional equation, that then become the algebraic models of curves and straight lines. Unless bounded in some way, such sets are infinite in enumeration.

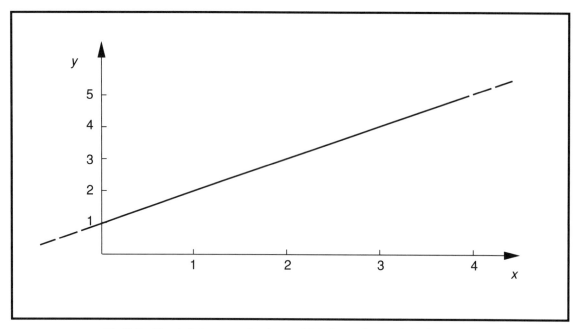

Fig 5.8   The infinite set of points which form the straight line $y = 2x + 1$

Figure 5.8 shows part of a straight-line set:

LINE $= \{(x\ y)$ where $y = 2x + 1\}$

which is an infinite set generated by the rule $y = 2x + 1$. (Note that LINE does not need to be an ordered set.) A finite subset of this set—representing four points on the line—might be

POINTS $= \{(1\ 3),\ (2\ 5),\ (3\ 7),\ (4\ 9)\}$

[We could in fact write this in the same generating form by making the definition more explicit:

LINE $= \{(x\ y)$ where x and y are positive integers and y less than 10
and $y = 2x + 1\}$]

If we need the set to reflect the physical sequence along the line, then POINTS must be an ordered set:

O-POINTS $= ((1\ 3)\ (2\ 5)\ (3\ 7)\ (4\ 9))$

Three-dimensional algebra requires ordered-triples and space-time algebra ordered-quadruples—but relativity theory is beyond the scope of this book...!

Note that if the sequence of sets in a PRODUCT expression is significant the resultant set is a set of ordered pairs, or Cartesian Coordinates if that is the context. This is in fact known as a Cartesian Product. For example, consider two sets:

$$A = \{w,x\} \text{ and } B = \{y,z\}$$

Then

$$A \times B = \{(w,y), (w,z), (x,y), (x,z)\}$$

but

$$B \times A = \{(y,w), (y,x), (z,w), (z,x)\}$$

which are not identical sets. And if the results are required to be ordered sets then the method of selecting elements also becomes significant. Above we have taken the elements of the first set in turn (w and then x for $A \times B$, y and then z for $B \times A$) and paired them with the elements of the second, which has affected the order of listing the results. It is important that the actual implementation of the product in any given language be known to the user.

5.4.2   Ordered sets as lists

The term that has become standard usage for referring to the representation and manipulation of ordered sets is **list processing**. Lists are ordered sets, though within definitions of list-processing languages they acquire additional properties to that of merely having elements in a defined sequence, and the term should be reserved for this usage therefore.

List systems incorporate the important property that the elements of each list can themselves be lists. Computer science calls such a property **recursion** and the recursive nature of list processing lends it great computing power potential. A fundamental reason for this is that recursiveness enables consistent list-processing rules to operate throughout implicit tree structures. Figure 5.9 shows this feature schematically. Not only is the general topology of the structure represented but the specific layout of the tree is defined by virtue of the nested elements being lists and not merely sets.

At the 'bottom' of any list structure there must ultimately be elements which are not in themselves lists, but the basic data objects for which the structure has been set up. These elements are referred to as **atoms**. Even that is not the end of the story necessarily, because there is no reason why an atom should not be a data object which is itself a label for further lists, such as the column heading in a table of data. How such things are dealt with will depend upon the syntax and grammar rules of the list processing language.

5.4.3   Heads and tails

A standard feature of all list definitions is that there is a leading element, followed by a variable number of other elements. The leading element is

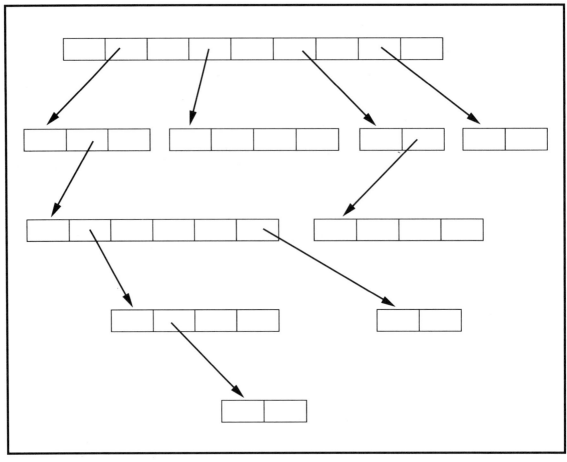

Fig 5.9 The recursiveness of the general list structure

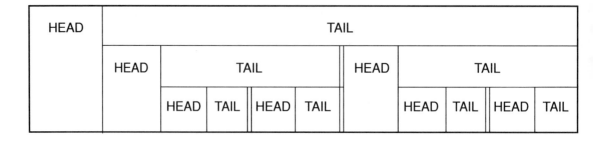

NB. This is purely schematic. The actual list structure will be managed by software in various ways using pointers, links, or indices.

Fig 5.10 Lists as heads and tails

given the name *head* (of the list) and all the remaining elements the *tail*. Thus there is no list 'body'—a list is sperm-like in comprising only head and tail! (The special case of the empty list can be dealt with by defining it as a list with empty head and empty tail.)

This feature applies throughout the list structure, so that lists nested within the main list are themselves deemed to comprise head and tail also (see Fig 5.10).

If a head is removed by an appropriate procedure, then the first element in the tail (the head of the tail!) becomes the new head. A list can never be headless. Likewise if a new element is inserted as the first in a list, it becomes the new head. A list can never be two-headed.

5.4.4 The LISP language

It is possible to construct lists in any programming language, though the task is greatly eased if the language incorporates specific list-processing functions. The LISP (List Processor) language does not merely incorporate such functions, its whole design is based upon list-processing principles. This gives it enormous strengths in respect of adaptability, flexibility and extendability, and, although it was originally invented in the late 1950s (by John McCarthy) predating even Cobol, it was not until the emergence of the demand for knowledge-handling languages in more recent times that it has blossomed into a widely available tool in computer-using environments. And not simply as a specialist knowledge-processing tool—some LISP environments may be found playing the role of comprehensive operating systems, for instance.

Apart from the usual command dialogue syntax rules, which we shall not be concerned with here, LISP is based on a single expression type—a remarkably simple notion in itself. Because LISP is seen at root as being a processor of the most fundamental atoms of data these expressions are known as *symbol-expressions* or just s-expressions. The simplest manifestation of the s-expression is as a single atom. But in general it is a list in the same sense which we have explained the term and therefore is recursive in nature (see Fig 5.11).

The layout syntax is the one we have already adopted, ie. round brackets and spaces as separators. However, the head of a LISP list is special in that it is taken to be a *function* and the tail is taken to be the *arguments* upon which that function is to operate. The number of arguments which the language-interpreting software expects to find depends upon the function and its definition. The language itself incorporates built-in functions which have defined numbers of arguments (either a fixed number, such as with the arithmetic functions, or a variable number, such as with list-handling functions). But LISP users can define their own functions (using the built-in functions, naturally), and therein lies the secret of the extendability of the language. Function definition is not unique to LISP by any means but,

```
S-EXPRESSION   =   NUMBER  ATOM

                        or

                   SYMBOL  ATOM

                        or

               (LIST  OF  S-EXPRESSIONS)
```

Fig 5.11   The composition of a LISP s-expression

combined with the fact that functions are themsleves part of the list structure, the feature proves to be very powerful.

To complete the picture, the arguments can, according to the nature of the associated functions, be anything from atoms to complexly structured lists. So while all the usual language functions are available at the simplest 'flat' tree level, without change of general form, all the specialised list-processing functions are available too—and anywhere into this can be inserted the user-defined functions suited to the particular application.

The number of built-in functions is well in excess of 150 in contemporary versions of LISP, and it is not the intention here to teach the language. But a tableau of the more easily understood functions is given in Fig 5.12, for the reader to browse through. In some cases the clarity of the function coding leaves much to be desired and it is possible that many would-be LISP learners have been put off very early by car and cdr and such like. In fact, these codes are a reflection of LISP's true age: car is an abbreviation for content address register; and cdr an abbreviation for content decrement register; terms meaningful only to users of low-level (machine-related) programming languages and even for them not necessarily regarded as being well-chosen explanatory abbreviations!

It is worth looking a little closer at the LISP conjunctive (and) and disjunctive (or) functions here, because they help to illustrate how standard logical concepts can be given special operational features in a given practical context.

They are both what LISP calls predicate functions in that when evaluated they can, if appropriate, return a TRUE (t in LISP) or FALSE (nil in LISP) result. The argument expressions themselves might also be predicate types, eg. (null 'list) which tests whether list is empty or not and returns t if it is and nil if it isn't.

The predicate and evaluates each expression, working from left to right down the list, until it finds a nil-returning expression, when it stops any further evaluation and returns the value nil. This reflects the rule that only one clause in a conjunctive expression need be FALSE to make the whole

In the panel below, the E characters symbolise any LISP s-expressions. Note that the first element in each LISP list (within the brackets) is always the function name.

(and El E2 ... En)	Evaluates the expressions from left to right and returns nil (for FALSE) as soon as a nil expression is found. Returns the value of En if none are nil. Returns t (for TRUE) if no expressions are supplied at all! Thus it tests for the presence of at least one nil expression.
(append El E2)	Returns a new list comprising the elements of El and E2 in that sequence.
(atom El)	Returns t if El evaluates as a single atom and nil otherwise. Thus it tests if the expression is an atom or not.
(car El)	Returns the head of El.
(cdr El)	Returns the tail of El.
(equal El E2)	Returns t if El and E2 are identical and nil otherwise.
(length El)	Returns the count value of the number of elements in El.
(not El)	Returns t if El is nil and otherwise returns nil. Thus it is a test for an empty list.
(or El E2 ... E3)	Evaluates the expressions from left to right and if all are nil returns nil. Otherwise returns the value of the first non-nil expression it finds. Returns nil if no expressions are supplied! Thus it tests for the presence of at least one non-nil expression.
(plus El E2)	Returns the sum of the (necessarily) numerical values of El and E2. Returns zero otherwise.
(quote El)	Returns the expression El, not its value! This is especially important where El is just a list of data elements, and not a LISP s-expression which must have a function name as its first element. (See Fig 5.13.)

Fig 5.12 Some of the LISP functions

expression false. If given no arguments at all, and returns t, reflecting that nothing has been shown to be FALSE. But if there are expressions as arguments and none of them evaluate to nil the predicate returns the value of the last expression in the list. This might be t if it is a predicate function but could be some other result such as a calculated numeric value, a character-string, or a list resulting from some list process. So, in spite of

141

being an apparently purely logical connective, and can return values other than TRUE or FALSE.

The predicate or likewise evaluates from left to right and stops evaluating as soon as an expression returns a non-nil result, when it will return that non-nil result itself. If all results are nil (or there are no arguments to evaluate) or returns nil itself. This reflects the rule that only one clause in a disjunctive expression need be TRUE to make the whole expression TRUE, except that here TRUE might be something other than the Boolean TRUE.

The predicate not is less surprising in its operation. It returns t if its (single-expression) argument evaluates to nil and otherwise returns nil. The only oddity here is that it returns nil for any 'successfully' evaluating expression rather than just for one that evaluates to t.

A tableau of simple examples is given in Fig 5.13.

These logic-related functions may seem 'impure' to the logician, but an experienced LISP programmer will readily avow to the functional effectiveness of these LISP implementations of the logical operators.

(equal 2 2) Returns t since 2 is indeed equal to 2!
(equal name (quote jones)) Returns t only if the list name comprises the single element jones. Equivalent to asking "Is the name Jones?"
(and (equal dob 1937) (equal name (quote jones))) Returns t only if dob (date-of-birth) evaluates to 1937 and the name is Jones (as per previous example).
(or (equal 2 5) (sum 1937 1) (equal name (quote jones))) Returns 1938 since a non-nil list is encountered as the second expression and duly evaluated.
(equal (car (quote (alwyn jones))) (cdr (quote (jones alwyn)))) Returns t since alwyn is indeed the head of the list alwyn jones and the tail of the list jones alwyn.
(length (append (quote a b) (quote c d e))) Returns 5, this being the number of elements in the list created by adding c, d and e to a and b.
(not (or (equal 2 3) (equal 3 2))) Returns t, since 2 does not equal 3 (nor does 3 equal 2!) and this nil return is then reversed to t by the not function.

Fig 5.13 Some simple examples in LISP

5.5 Relations

5.5.1 Introduction

Relational concepts have become a very familiar part of the database scene and most systems analysts will already have met them in that context. We shall make it our aim here to attempt to explain how the relational concepts fit into the picture we have developed of sets and lists. It is not intended to present an overview of, or even an introduction to, the practical aspects of relational database design and use (there are plenty of good texts that do this very well), but understanding the context should help the reader to understand how the relational database languages are really part of a spectrum of approaches to the general question: "How can set and list ideas be used as a basis for practical data and knowledge processing?".

Although relational data-handling principles were developed from mathematical theory (initially by E. F. Codd in the early 70s) it is undoubtedly true that they are easier to understand initially from a practical viewpoint of data-handling issues themselves. We can all readily appreciate the structure and use of general tables of data, and it is via such structures that the meaning and purpose of relations can be made apparent. This therefore is the approach we adopt here.

5.5.2 From tables to relations

A table of data—we are talking here of straightforward tables, not decision tables—comprises, if arranged vertically, a number of headed columns with rows of entered data in vertical alignment below these heads (see Fig 5.14). Such tables are ubiquitously and perennially useful display devices, but it is important to realise at the outset that they provide both the advantage of standardising the data itself (referred to as *normalising* as we get more formal) and the disadvantage that data is thus compartmentalised and shoehorned into formats that are not altogether natural to the way, as intelligent beings, we handle data and, more especially, knowledge. However, the disadvantage is something to return to later, it is the advantages that we shall be highlighting here.

TABLE HEADING

COLUMN HEADING	COLUMN HEADING	COLUMN HEADING	. . .
Data item Data item Data item . . .	Data item Data item Data item . . .	Data item Data item Data item . . .	

Fig 5.14 A conventional table of related data

It is certainly not difficult to see that a table comprises sets of things. There is a set of column heads, sets of data in columns, and also sets of data in rows. The column and the row sets are not disjoint; indeed taken as a total set of sets they would seem to comprise identical content—all the data in the total table. Two questions should come immediately to mind in the context of this view: how significant is the sequence of columns left-to-right and how significant the sequence of rows top-to-bottom.

In setting up an actual table of data a writer or statistician may give careful thought to the arrangement of columns. The leftmost column is often assigned the role of indexing—it is anticipated that the table user will approach it with a known item of data from this column with the intention of looking up the data associated with it along the same row. For the same reason this column may show the data in a recognisable sequence—numerical, alphabetic, temporal, etc. However, these decisions are a matter of design choice. Different column arrangements and different (or no particular) sequences of data top-to-bottom are logically possible.

What is not alterable is the column to which an item of data is assigned once the heads and the data to be associated with it in the same row are in place. This is the essential nature of tabular presentation and it is this that is captured in the concept of the data relational form. So what have we got?

1 A set (not a list) comprising a definite number of column heads as elements, the set as a whole having an associated label which is the table name or title. This is a **relation**. A relation is therefore equivalent to the framework for a whole table of data.

2 Each column is a label for a set of data values, the number of which depends upon circumstance. Let us call these **data sets**. Many other names are in use, especially in different contexts such as database theory. We shall make use of these as appropriate, eg. *attribute*, *field*, *data-item type*.

3 Each row of the table is a set comprising the same definite number of elements as the columns mentioned in para 1, each element being an actual data item which must be a member of the appropriate data-set. This row set is termed a **tuple** and the whole relation is described as being an n-tuple relation, where n is this definite number of data-sets (column heads). The data items in each tuple must form a list because they are in a sequence determined by whatever is the sequence of the data-set names.

4 The body of the table, comprising all the data items, is thus also a **set of tuples**. The sequence of the tuples is immaterial and there can be no duplicates. The tuples have no individual labels (there are no row references in the relational type of table), their identity residing in the uniqueness of each list.

5 In order to avoid ambiguities it is essential that every item in every row be understood as explicitly entered. This is in contrast to informal tables where blanks or dashes may be used to imply 'ditto'. The ditto relies upon rows being in ordered groups of some kind and we have already stipulated in para 4 that the sequence of tuples in a relation is undefined. (This doesn't stop a relation from being sorted, of course, in a practical case. But once sorted it is then a special kind of database relation.)

Taking all these together and summarising, we have therefore: a relation name (table name), data-set names (column heads), and unlabelled tuples (table rows) holding the data items. One convention is to write this as

RELATION-NAME(Data-set-name, Data-set-name,...)

when representing the relation as a whole, and to write

(Data-value, Data-value,...)

in the correct (ie. currently displayed) column-head sequence, or more explicitly:

Data-set-name	Data-set-name	...
Data-value	Data-value	...

when representing actual tuples. (Conventions about the use of upper and lower case for relation and data-set names do vary, but we shall use them as shown.)

The relational notation, while clearly bearing an appropriate family resemblance to that adopted for sets, has the special feature of showing explicitly the relation name together with its set element names. The data tuples are sets constrained to be lists by the selected format of the relation to which they belong. It is a notation now widely accepted and, though dressed up in different ways with respect to the characters and separators, appears in many computer data manipulation languages.

We could have arrived at this point by formally (and much more rigorously) defining the relational format as well-formed mathematical formulae. Or we could have taken an actual relational data manipulation language and dissected it to find out how it works. We chose neither of these routes because we wanted to show how the relational form is a bridge between the theory of sets and practical data manipulation considerations. We believe that it is a bridge which the systems analyst should be able to cross in either direction. (There will be a closely related bridge when we come to look at predicate logic in Chapter 8.)

Finally, the correspondence between relations and traditional data processing files is worth confirming since many longer-experienced systems analysts will be as familiar with the latter as with anything when it comes to computer handling of data. The correspondences are as follows:

Relation: unsorted File containing single record-type
Relation-name: File label
Entity set: the format of the File record
Tuple: an actual record on the File.

But the correspondences are conceptual only since conventional data processing files have usually been created by a specific programming environment (eg. Cobol) that processes the files in thoroughly idiosyncratic ways which may bear little resemblance to operations with sets. Such arrangements may be satisfactorily efficient in their own terms, but start to present serious problems if such files need to be drawn into a wider knowledge environment. Special programs to transform or translate the files into more universally recognisable set-like formats will almost certainly be necessary in such circumstances. What has begun to happen in practice is that the wider environment has arrived in the shape of database-oriented software systems (eg. Oracle) which, while being closer to set-theoretic forms, still demand special formats related to their own data handling rules.

Although this situation may or not be satisfactory in terms of operational efficiency and effectiveness, it is desirable that the systems analyst be aware of the broader picture, especially in the context of any knowledge processing requirements.

5.5.3 Working with relations

However, because of the special set structures inherent in relations, the operations on relations need to be rather more than the standard set operations. And tuples are not standard list forms either, so any list processing requirements will also be different. To consider what is appropriate let us return to the data table analogy, which is after all a very accurate analogy as we have seen.

First we must make a general rule that enables the relation manipulation to be more easily automated. This is simply that every operation on a relation or a number of relations results in the production of a new relation. Even if the outcome is a single data item, in relational terms this is a relation with one data-set and only one row! In more formal language terms:

(new-relation-name)
= OPERATION performed upon (relation-names)

We shall now proceed to look briefly at four such operation types, which should form a core of all relationally oriented languages, even though the actual syntax and grammar will vary from one language to another.

5.5.4 Tuple selection operations

When we have data before us set out as tables, how might we make use of it? The most obvious case is that we might wish to find a specific data item (or group of items) in order to look at the associated row data. Therefore an operation TUPLE-SELECT is defined for relational working. But selection is always based on some criterion or criteria, so the TUPLE-SELECT operation has to be accompanied by a criterion clause. The general term for such a clause is **predicate**, which we have already had cause to mention as being a very important term within knowledge processing in this and other contexts. The syntax used for signalling the predicate varies with computer language. For clarity of meaning we shall use a prepositional signal WHERE. Thus the complete format for selection is

(new-relation-name)
= TUPLE-SELECT (relation-name) WHERE (predicate)

The relation whose name is given at (new-relation-name) will comprise the same data-sets as (relation-name) but its tuples will be a subset of the tuples of that relation. The selected rows are copied out, as it were, into the new relation, leaving the subject relation in an unaltered state.

The rules for writing out the predicate will depend upon the language in use, but the kinds of criterion that can be specified are, for example:

a Equality with a data item, for example:
employee-no = "22564"
price = 24.50
product-name = "door-bolt"
reorder-flag = 1, ie. a binary TRUE

b Value-range for a data item, for example:
employee-no < "40000"
price > 50.00 < 150.00
town-name > "Widnes" < "Yarmouth"
  ie. between the two in alphabetic sequence.

c Tests on sets of values, for example:
All(amount) < 100, ie. are all amounts less than 100?
Any(amount) < 100, ie. is there any amount less than 100?
Exists(amount), ie. is there any 'amount' item?

Note that the above are all Boolean tests—they are each either TRUE or FALSE. (Note the last line under **a** is an example of a Boolean test of a Boolean variable—there are two levels of truth implied!) Therefore we can also have

d Boolean combinations of the above, for example:
NOT(factory = "London" OR "York") AND (price < 75.00
  OR stock > 1000).

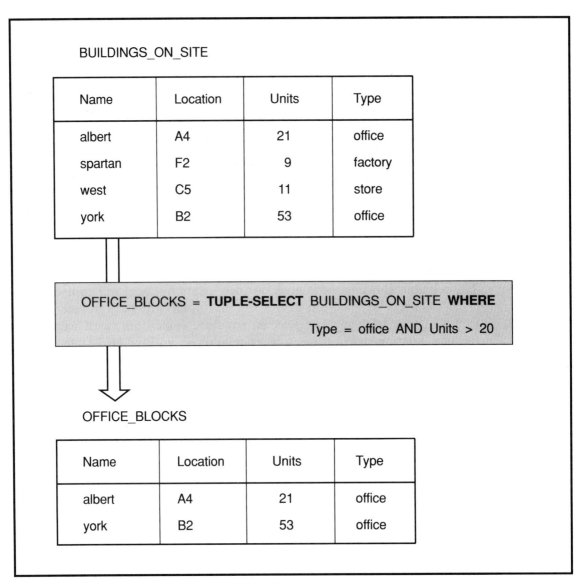

BUILDINGS_ON_SITE

Name	Location	Units	Type
albert	A4	21	office
spartan	F2	9	factory
west	C5	11	store
york	B2	53	office

OFFICE_BLOCKS = **TUPLE-SELECT** BUILDINGS_ON_SITE **WHERE**

Type = office AND Units > 20

OFFICE_BLOCKS

Name	Location	Units	Type
albert	A4	21	office
york	B2	53	office

Fig 5.15 An example of a TUPLE-SELECT operation

Learning the syntax and grammatical rules for writing out predicates represents a large proportion of the difficulty in becoming adept at using a data manipulation language, as many systems analysts will know only too well.

An example of a TUPLE-SELECT operation is shown in Fig 5.15.

5.5.5 Set operations UNION and INTERSECTION

Suppose we have two tables of data from different sources which are about the same things. They have exactly the same column heads but the body of

148

data is different, though there may be some tuples from each which are identical to one other. It is not unlikely that we would want to combine these into one table, dropping the duplicates in the process.

This is the UNION operation

> (new-relation-name)
> = (relation-1-name) UNION (relation-2-name)

This is entirely consistent in operation with the definition of the union of sets. With relations, since there is no significance in the tuple sequences, the operation is not a merge of the two sets as is the case when bringing together data from two files in data processing. There is no stipulation about tuple sequence in the result, though the most straightforward action in practice is to append one table after another, dropping the duplicates from the appended table.

An example of a UNION operation is shown in Fig 5.16.

If, on the other hand, we wanted specifically to isolate the duplicates we would be looking for the standard INTERSECTION of two sets. This is then:

> (new-relation-name)
> = (relation-1-name) INTERSECTION (relation-2-name)

An example is shown in Fig 5.17. The two subject relations must in each case be compatible with respect to data-set (same column heads) and any implementation of these operations must stipulate how non-compatibility is handled if encountered. For a comment about synonomous names for data-sets, see under JOIN below.

5.5.6 PROJECTION operations

Another way in which we may want to extract data from a table is to focus on data in certain columns, the data in the other columns being discardable. PROJECTION is defined to achieve this.

In this case the predicate must be quite specific; it must cite a set of data-set names to be retained (or, in a closely related form of the operation, of those to be discarded). We adopt here the preposition ON to signal this. We thus have:

> (new-relation-name)
> = PROJECT (relation-name) ON (data-set1, data-set2,...)

The number of tuples in both sets will be the same, but each tuple in (new-relation-name) will comprise the reduced list of data items.

TUPLE-SELECT and PROJECT can be used one after the other to arrive at a single-item relation or one comprising a small block of items. Note that if PROJECT is used first the possible selection criteria are reduced. In any case, whether it is more efficient to use TUPLE-SELECT first or PROJECT first will depend upon the dimensions of the relations, the

BUILDINGS_ON_SITE_ONE

Name	Location	Units	Type
albert	A4	21	office
spartan	F2	9	factory
west	C5	11	store
york	B2	53	office

BUILDINGS_ON_SITE_TWO

Name	Location	Units	Type
askey	A1	2	shop
spartan	F2	9	factory
west	C5	11	store
yarrow	A4	21	office

BUILDINGS = BUILDINGS_ON_SITE_ONE **UNION** BUILDINGS_ON_SITE_TWO

BUILDINGS

Name	Location	Units	Type
albert	A4	21	office
spartan	F2	9	factory
west	C5	11	store
york	B2	53	office
askey	A1	2	shop
yarrow	A4	21	office

Fig 5.16 Example of a UNION operation

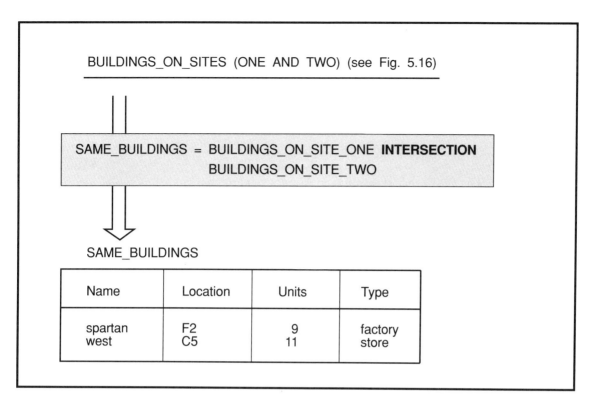

BUILDINGS_ON_SITES (ONE AND TWO) (see Fig. 5.16)

SAME_BUILDINGS = BUILDINGS_ON_SITE_ONE **INTERSECTION** BUILDINGS_ON_SITE_TWO

SAME_BUILDINGS

Name	Location	Units	Type
spartan	F2	9	factory
west	C5	11	store

Fig 5.17 Example of an INTERSECTION operation

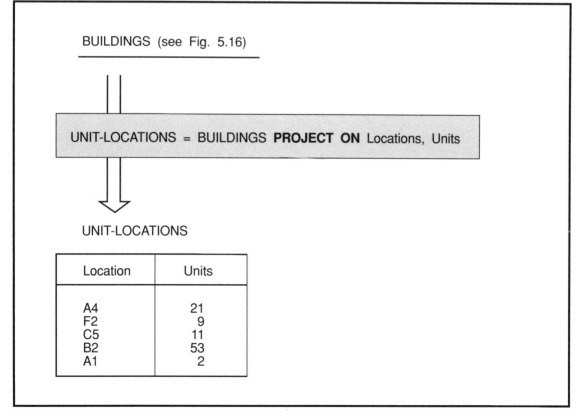

BUILDINGS (see Fig. 5.16)

UNIT-LOCATIONS = BUILDINGS **PROJECT ON** Locations, Units

UNIT-LOCATIONS

Location	Units
A4	21
F2	9
C5	11
B2	53
A1	2

Fig 5.18 Example of a PROJECT operation

breadth of the selection, the degree of stripping of the projection and, probably the most important of all, the performance characteristics of the software carrying out these operations.

An example of a PROJECT operation is shown in Fig 5.18.

5.5.7 JOIN operations

The JOIN operation combines data from two relations which do not have identical data-sets but do have at least one data-set in common. The common data-sets are used as criteria for forming the resultant relation. We shall use the preposition TO between the relations to indicate a 'first' and 'second' subject relation:

(new-relation-name)
= JOIN (relation-1-name) TO (relation-2-name)

The phrase 'data-set in common' is easiest to understand when this means an identical column head in each table. But it is possible that actual column heads are synonyms for the data-set name—indeed some implementations may insist upon this by not allowing identical names to appear in relations! It is advisable to use 'family names' such as Part_No_A, Part_No_B, where such constraints apply. The problem when synonyms are not allowed is that the user has to keep tight control on ensuring identical names throughout.

Synonomous data-set names give rise to the idea of there being a generic entity (Part_No in the above example) and the question then arises as to what data elements may be in the generic data-set. This is sometimes referred to as a **domain** of the data. But the actual extent of a domain is quite a troublesome issue to nail down and we shall come back to it when we discuss predicate logic in Chapter 8.

If a pair of data tables have at least one data domain in common (but not all) there are various ways in which any corresponding values in the two tables can be used to bring the data together. The simplest situation to envisage is where one table is actually a selection-set criterion for the second. Such a table would comprise just one or more column heads in common with the other (but contain none that were not in it). The effect is then just like a TUPLE-SELECT operation. See Fig 5.19.

But supposing there are data-sets in both tables not common to the two of them. We can still regard the first as a tuple-selection set for the second, but what about the other data in its tuples? JOIN can be defined so that this data is added to the resultant relation, which then has a data-set which is a union of the two subject data-sets. The result is therefore like a table showing all the data related to any item which is common to both tables. And if the data item is repeated in one or both of the tables new tuples are generated for each occurrence of it. See Fig 5.20.

When there are two or more columns in common, the question arises as

152

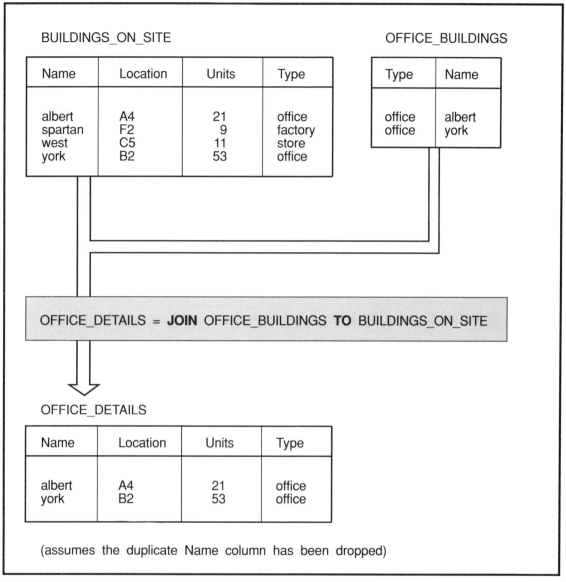

Name	Location	Units	Type
albert	A4	21	office
spartan	F2	9	factory
west	C5	11	store
york	B2	53	office

OFFICE_BUILDINGS

Type	Name
office	albert
office	york

OFFICE_DETAILS = **JOIN** OFFICE_BUILDINGS **TO** BUILDINGS_ON_SITE

OFFICE_DETAILS

Name	Location	Units	Type
albert	A4	21	office
york	B2	53	office

(assumes the duplicate Name column has been dropped)

Fig 5.19   Example of JOIN as TUPLE-SELECT

to whether the data should be joined only if there is a total match. If joining takes place on partial matches the result can be voluminous and confusing. In practice, different variations of JOIN have to be defined with slight variations in syntax in order to distinguish between the different functions.

5.5.8   From the algebra to a language—a brief look at SQL

What we have been presenting over the last few sections is a *relational algebra*. What is needed in a practical data handling environment is a

153

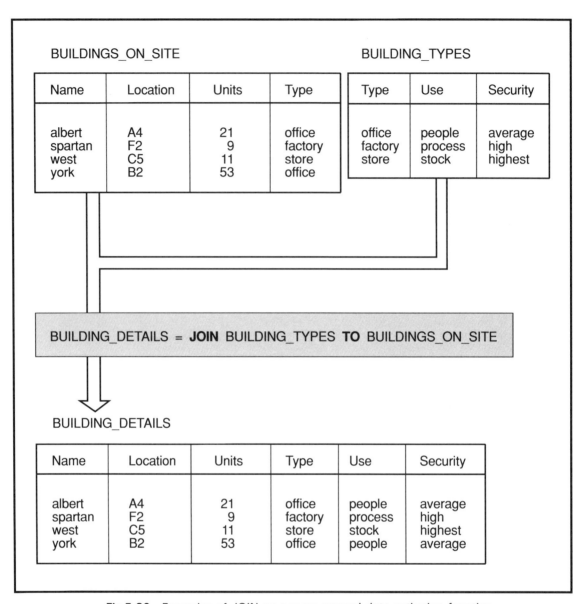

BUILDINGS_ON_SITE

Name	Location	Units	Type
albert	A4	21	office
spartan	F2	9	factory
west	C5	11	store
york	B2	53	office

BUILDING_TYPES

Type	Use	Security
office	people	average
factory	process	high
store	stock	highest

BUILDING_DETAILS = **JOIN** BUILDING_TYPES **TO** BUILDINGS_ON_SITE

BUILDING_DETAILS

Name	Location	Units	Type	Use	Security
albert	A4	21	office	people	average
spartan	F2	9	factory	process	high
west	C5	11	store	stock	highest
york	B2	53	office	people	average

Fig 5.20 Examples of JOIN as a more general data-gathering function

meaningful language tool which enables the user to enjoy the benefits of the algebra transparently. A method of implementing an algebra as a set of procedural and grammatical tools is called a *calculus* by mathematicians and computer scientists, but 'algebra' and 'calculus' are relative rather than absolute terms and there is really little need for the non-purist to agonise over the distinction. (See Section 2.5.1.)

We choose SQL as our illustrative model not simply because it has evolved (indeed is still evolving) quite genuinely from the algebraic theory but also because it is showing every sign of becoming a *de facto* industry standard—though a defined standard version of it is still not quite with us at the time of writing. It must be pointed out that there are many other *relational languages*, some more worthy of the epithet than others, but the reader is directed to other texts for reviews and comparisons of these. It is hoped that the review of the SQL implementations of the algebra, and of some functions which appear to go beyond the algebra, will provide a useful introduction to other more technical studies.

We shall not be attempting a full introduction to SQL with all its data handling functionality. We shall focus on the ways in which the algebra has been given syntactical clothing, rather than, for instance, on how the user gets into interactive mode with the language.

The basic data relation in SQL is called a *table* and in the first instance we need say no more about it than that. The first thing to grasp about SQL is that it is typical of the approach which takes the embarrassment of choice about operational sequence away from the user. It does this by providing one main type of sentence structure for table manipulation, called the SELECT statement, which comprises six *clauses* (in the grammatical sense), most of which are optional, but whose sequence of appearance is constrained and whose sequence of action on the data is predetermined. It generates (un-named) *intermediate result tables* from each stage of the interpretation of the statement and the *final result table* is a screen display (still un-named; to give results names the user makes use of a CREATE VIEW statement, but we shall not be concerned with such issues here.)

The SELECT statement signals the object table (or tables) to be worked upon by use of the clause FROM. The simplest form of the SELECT statement is in using it to display the whole of a table. This is

```
SELECT *
FROM (TABLE-NAME)
```

Projection is achieved by replacing the asterisk (which formally represents 'all columns') with a list of table-column names (data-set names), which must of course be a subset of the column names in the object table. So, projection is

```
SELECT (Column-name, Column-name,...)
FROM (TABLE-NAME)
```

Projection is achieved therefore without an explicit verb, the position immediately after the SELECT verb being the signal instead. Tuple selection is achieved by following the above with a WHERE clause which predicates the selection. This is

```
SELECT (Column-name, Column-name...)
FROM (TABLE-NAME)
WHERE (predicate)
```

The predicate can be built up into a Boolean expression using AND, OR and NOT on the conditions as discussed in Section 5.5.4. But SQL greatly extends the power of predicate writing in a number of ways. The most notable of such extensions is provided by allowing the SELECT clause itself to be used within the predicate in what is termed a *subquery*. A subquery itself can carry a further subquery, and so on. Thus test conditions can be applied to sets of data (sub-tables) which are intermediate result tables at many levels.

Another important SQL feature is that it allows for *null-valued data*, ie. a data item where a value has not been specified. Thus a Boolean test can be made as to whether an item is null or not by use of a NULL test operator. This returns TRUE or FALSE, of course, but there are other test functions which, when they encounter null data, return a NULL result, which therefore has to be defined within algebraic operations together with TRUE and FALSE. (This means that SQL expressions are three-valued and therefore not truly Boolean!) Table 5.1 shows the definition.

While one TRUE is sufficient to make {A OR B} TRUE and one FALSE is sufficient to make {A AND B} FALSE as one should expect for consistency with Boolean definitions, the NULL otherwise 'contaminates' the expressions and gives a NULL result overall. The NULL value can be thought of as 'cannot be decided' in this context, though it has rather more connotation than that. It is certainly necessary to keep a clear distinction in mind between the test for null-ness, which is Boolean, and the returned result NULL which is not. Naturally, this is largely a matter of learning the language skills by actually using SQL.

There are number of SQL operators for dealing with sets within predicates. Here are the main ones.

The IN operator enables set membership to be tested, by

WHERE data-name IN set-name

This is TRUE for every instance where the named item is a member of the named set and is actually equivalent to the disjunction of equality tests throughout the set members, ie.

WHERE (data-name = member-1) OR (data-name = member-2) OR...

Table 5.1 *How the logic connectives work in SQL*

A	B	A AND B	A OR B	NOT A
TRUE	NULL	NULL	TRUE	FALSE
FALSE	NULL	FALSE	NULL	TRUE
NULL	NULL	NULL	NULL	NULL

The set-name can of course be replaced with a subquery. Also, the IN operator can be preceded by NOT to test for non-membership.

The ANY operator enables a comparison test, rather than membership test, to be made against a whole set. For example,

WHERE data-name > 25 ANY set-name

The set-name must of course comprise the same data type as data-name for this to make sense. The test is then TRUE if any of the occurrences of the named data in the set passes the test (>25 in this case). It is equivalent to

WHERE (member-1 > 25) OR (member-2 > 25) OR ...

The ALL operator is similar except that the test must pass for every member of the set. Thus,

WHERE data-name > 25 ALL set-name

is equivalent to

WHERE (member-1 > 25) AND (member-2 > 25) AND ...

The UNION and INTERSECTION of tables can be achieved in SQL very straightforwardly by linking SELECT statements with a UNION and INTERSECT operator respectively. The result table of each SELECT statement must comprise the same columns of course so that the result of UNION will be the rows of one table appended to the rows of the other and of an INTERSECT will be rows common to both result tables. (But note that since INTERSECT can be achieved quite easily using an appropriate WHERE predicate it is not yet part of standard SQL, though it is to be found in some dialects.)

The JOIN of SQL tables has to be achieved by the appropriate use of the WHERE clause. If two tables are cited after the FROM clause, and the WHERE clause uses a predicate relating data in one table to data in the other, then a JOIN of the two tables will result. For instance,

```
SELECT Column-name list
FROM Table-1, Table-2
WHERE Table-1-data-name = Table-2-data-name
```

achieves a standard JOIN of the two tables (called an EQUIJOIN). Other types of JOIN are achieved by combining variations of the WHERE predicate with the use of the UNION operator. This throws the onus on constructing the manipulation of instruction on to the user and the task reverts to being more algebra-like than calculus-like.

We have still only mentioned three of the six possible clauses in a SELECT statement. The remaining three are to do with the arrangement of rows within tables, in the sense of top-to-bottom sequence. They are, in the

Table 5.2 *The interpretation sequence of SQL clauses*

Stage	Clause	Action
1st	FROM	Takes the appropriate table(s)
2nd	WHERE	Makes the appropriate row selection
3rd	GROUP BY	Forms rows into groups
4th	HAVING	Further selects rows within groupings
5th	SELECT	Projects the appropriate columns
6th	ORDER BY	Sorts rows into final presentation sequence

required sequence:

    GROUP BY ...
    HAVING ...
    ORDER BY ...

The last is also the last to be performed by the SQL interpreting software and specifies the row presentation sequence of the final result table. GROUP BY, followed optionally by HAVING, is carried out before the final PROJECTION (the column names cited after SELECT) and gathers together rows with matching criteria as specified after GROUP BY. These grouped sets of rows can then be subjected to further processing by use of HAVING, which in many respects is the WHERE clause in a different context.

Although we have not here been concerned with practical implementation matters we present in Table 5.2 the sequence in which the clauses of a SELECT statement are interpreted and acted upon since this does show how the relational algebra is acted out in SQL. Any subqueries will have similar interpretive sequences embedded within WHERE.

There are many who argue that SQL is too far from being the ideal of a good-quality calculus for a 'natural' data handling language based on the algebra to be taken up as a standard as yet—hence the epithet '*de facto* standard', although purveyors of competing languages would dispute even that. It did however suit our purpose here—that of examining how the algebraic ideas might be implemented in an actual language.

6 Decision Logic

6.1 Introduction

The making of a decision is an intelligent activity worthy of considerable study in its own right. The human decision maker is inevitably using a vast amount of data from a wide range of types when making even the simplest of decisions. A seemingly purely 'physical' decision such as "I shall close my mouth" may be made on the basis of extremely complex social and physical environmental conditions—although it is true that the decision maker may explain the action in relatively simplistic terms such as: "because I wanted to" or "because I had nothing to say" or "in order to avoid swallowing flies". But such surface simplicity conceals a considerable amount of secondary contingent information concerning state of mind, intention or need to control. How much more complex then is the implicit structure behind a decision such as "Sell tin" in a busy commodity market.

Figure 6.1 shows this situation as an input–output model. As far as our wish for a logical perspective is concerned, the focus is on the central box

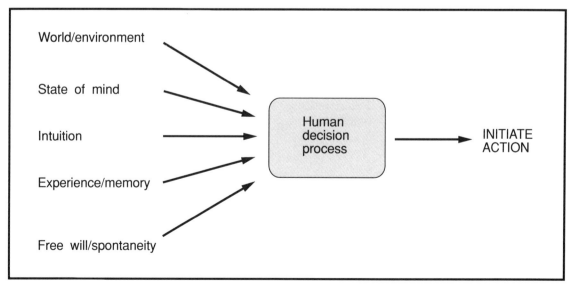

Fig 6.1 Some of the factors in human decision making

159

only, and only input which is specifiable as factual statements will be considered admissable. We shall not therefore attempt to engage in the study of what might be called the psychology of decision-making—or decision-taking as it might be more aptly termed in that broader context.

The logic must concern itself only with the processes of translating factual conditions into an appropriate prescribed action. This phrase implies a pre-existing knowledge base of appropriate actions which are prescribed for sets of expected conditions. This is of course extremely problematical in practice—how do we generate lists of expected conditions, let alone face the question of finding the right action to suit all the possible combinations of those conditions? In broad terms this is the challenge of knowledge acquisition theory and technique and again is not to be our concern here. Thus we have simplified our frame of interest to the model shown in Fig 6.2.

We assume a feasible construct whereby the decision-making process is essentially a question of receiving a set of known conditions defining the state of the real-world system, feeding these to a conditions-to-actions knowledge base and thereby generating a prescribed action (or actions). It is a crude model, but one that we shall be able to develop in sophistication as our understanding of modelling techniques advances. For now we shall use the term *decision logic* to mean this simple model.

The structural modelling of decision logic has been important in many branches of applied mathematics and science for some considerable time. Combinatorial and probabilistic analysis of decision sets has been widely applied in business and industry over a range of activities stretching from human problem-solving to factory process control. And the programmable computer has provided ample opportunity to capture decision structures for iterative use with different input values.

More recently, interest in decision logic has been boosted by the development of ideas related to knowledge processing, especially in the field

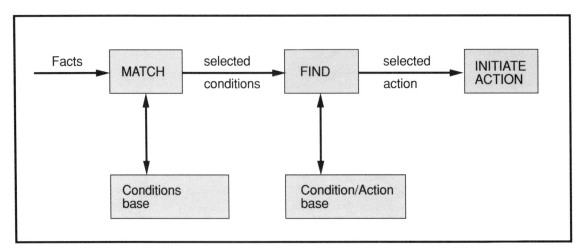

Fig 6.2 A possible framework for a simple facts-to-action process

of expert systems design. It has become increasingly obvious that modelling techniques needed to be developed with very much more variety and adaptability than has been evident in the field so far. However, the aim of this chapter will be limited to a review of the ways in which the structure of decision logic can be usefully modelled.

6.2 Representations of decision logic

6.2.1 The truth table as a representation

Truth tables as we have seen in earlier chapters enable decision logic to be represented when the following situation applies:

a) All condition variables are binary, ie. the condition can only be true or false.
b) The condition set is complete.
c) For every possible permutation of all condition set values the required output is known in the form of the truth or falsehood of a given function, ie. the function is also a binary variable

What kind of real-world problem situation meets these requirements? Well, binary logic circuits should! It should be possible to know what the circuit will output for every possible combination of input signals. For example, a circuit might be designed to cycle the pattern of bits in a register. The inputs would need to be

a) A current bit-state for each bit-position in the register.
b) A signal indicating the direction of cycling required, right or left.
c) A bit-set containing the numerical count of the number of bit places required to shift.

The output would be

The new current bit-state for each bit-position in the register.

It is perfectly possible to devise a truth table to fully specify this function, though there may still be questions to be answered about what state *b* and *c* input sets should be in after the function has been performed. A sample operation of the function is shown in Fig 6.3.

Let us imagine that we train a human operator to perform this function so that the process could be said to comprise human decision-logic. If we required the function to be performed at speed and over long work periods it is certain that the human operator would be very prone to making errors. But we can just about imagine the situation as a real one and therefore justify our description of the circuit as a decision-logic processor.

This is important to us. Imagine that we encountered an operator actually performing this task. As systems analysts it would be valuable to us to feel

161

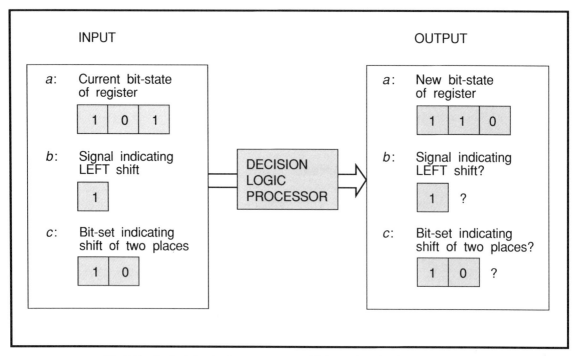

INPUT

a: Current bit-state of register

| 1 | 0 | 1 |

b: Signal indicating LEFT shift

| 1 |

c: Bit-set indicating shift of two places

| 1 | 0 |

DECISION LOGIC PROCESSOR

OUTPUT

a: New bit-state of register

| 1 | 1 | 0 |

b: Signal indicating LEFT shift?

| 1 | ?

c: Bit-set indicating shift of two places?

| 1 | 0 | ?

Fig 6.3 A decision-logic processor which cycles the bits around a three-bit register

that we could expect to be able to fully specify the decision logic in such a way that the task could be redesigned as a circuit! In the past there have been many form-filling tasks carried out by clerical workers which were not too dissimilar to this situation. Systems analysts have more or less successfully carried out computerising of such tasks in the past, but many tasks are not of course so simple and long swathes of coded program instructions have been created, rather than solitary 'circuits'.

However, it might help our understanding of more advanced knowledge processing to regard the clerks as knowledge workers and experts, albeit in a very limited sense. The circuit discussed above represents a very rudimentary automated knowledge processor. Once we start to search for more sophisticated tools and machinery, we soon begin to realise the limitations of the truth table as a modeling device for the decision logic.

6.2.2 The decision table as a representation

We discussed in Chapter 4 how the decision table is in some respects merely a truth table differently set out to aid readability and usability. But in fact this difference in layout is indicative of much more than immediate convenience. The instinct to seek ways of enabling questions to be set out in more explicit form, rather than being represented by Boolean variables, is a first step in giving emphasis to the rules governing the conditions-to-actions transformation.

While the conditions are stated in the form of binary questions we are strongly reminded that the decision table is a better-dressed version of a truth table and this leads us to use what we see as a means of checking for completeness and rigour of the logic. But as we shall see, the binary question is not a strict requirement for decision table representation, a fact which is emphasised by the use of the prefix 'limited-entry' for such tables. Later, we shall consider the use of non-binary questions where the answers to them are 'non-limited' or 'extended' entry.

6.2.3 Tree representations

Tree diagrams and constructs carry direct visual information about sequences of decisions and imply paths through the structure from trunk, through branches to leaves. This information may be essential or merely significant or, weaker still, purely pragmatic. At worst it can be downright misleading. Consider the following examples:

- A tree-structure wall chart in a power-control room indicating the steps in shutting down the plant under various conditions. Here, the question sequences and path-branching according to observed reactions of control meters is likely to be essential to the proper and successful shutting down of the plant.
- A fault-checking chart, in a vehicle maintenance manual. It may be strongly significant that answers to more general questions be established first, such as "Engine not firing?," ahead of answers to more specific questions, such as "Strong smell of fuel?", but not absolutely essential.
- A guide chart to establishing eligibility for a grant or benefit. If the aim is to establish status with respect to a number of independent attributes and sub-attributes (marital status—if married, name of spouse; income level—if above £10 000 whether from more than one source; age—over 50, whether in receipt of pension, and so on), it may be of practical value to ask one main question ahead of another for reasons of simplicity of presentation, but there may be no logic to that sequence.

In the second and third examples, it is possible to misjudge the value of any particular decision structure and present ones which are misleading to their users.

These points illustrate both a strength and weakness of tree representations for modelling decision logic. The strength is in being able to express the hierarchical nature of the decision logic should it be there; the weakness is in requiring such a structure to be set out even if it is not inherent in the logic to be modelled.

A similar two-sided case relates to the open-endedness of the tree representation with respect to the conditions to be tested. There is no obvious evidence of exhausting the list of conditions as there is in the condition quadrant of a decision table, and furthermore it is very far from obvious whether all possible combinations of condition (ie. all possible paths) have been considered by the tree designer. This is a strength in providing flexibility for the designer, enabling just what is required and nothing else to be set out. The reader of the tree model is not distracted with explicit expression of many paths which lead to no action, or by many separate paths leading to error states. It is a weakness for the obvious reason that completeness of analysis is very difficult to establish where a rigorous examination of all possible cases is desirable.

6.2.4   Net representations

A network of decision paths implies the possibility of separate paths rejoining before the action end of the path is reached. It also opens up the

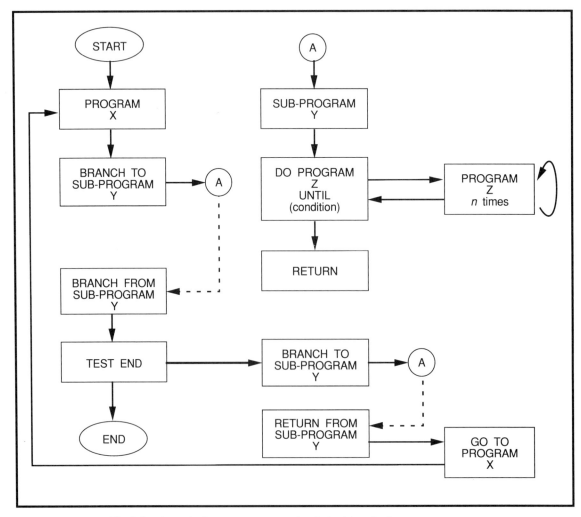

Fig 6.4   A typical program network of decision logic

possibility of a path re-joining itself and forming a loop. Conventional programming (up to the third generation of languages) certainly allows for these features through branching into sub-routines, iteration ('do until' and such like) and 'go to', though this last feature may be deliberately restricted in practice. See Fig 6.4 for a schematic diagram of such features.

The go to restriction just mentioned is indicative of the problems inherent in controlling the logic of unrestrained net structures. Any loop opens up the structural reality to the thorny question of time. If a loop section is in process of being used, is it for the first, second or what time? A controlled iteration will of course reveal the answer by virtue of whatever countdown is being used. A sub-routine will reveal such an answer only if a count is incorporated within the loop, usually an option rather than being in-built. The go to is the real culprit in this respect because it takes control to any part of the net in a completely anonymous way, carrying no clues as to where it came from and when. The go to that takes the control path into the middle of a sub-routine or iteration loop compounds the confusion. Hence the restricted use of go to in all self-respecting programming outfits. But structured programming theory recognises that it is important to do much more than control the use of go to if the net structures of program code are to be manageable—but this is an area we shall not 'go to' from here!

6.3   Decision tables in use

6.3.1   Building limited-entry decision tables

To build a decision table we must first have assembled the raw materials for doing so. This means having to hand some factual description of the decision logic existing in the real world situation we wish to model. It could be an existing set of instructions arising out of the documentation of business processes. It could be in tree or chart form. It might even be unrecorded as yet, being part of a worker's knowledge, experience or skill.

The availability of existing documentation would suggest a clerical, administrative or operative type of work. Unrecorded decision logic is much more likely to obtain in connection with the more skilled or 'expert' person. This latter situation is more the province of expert systems development. For now we merely observe that acquiring the understanding of the existing decision structure in some cases may well be far from easy. We shall give it the task-name:

Acquisition of knowledge about the decision logic

and say little about it here because our focus of interest is on the subsequent methodical building of the actual table.

To that first task-name we shall add nine more, to be followed more or

Fig 6.5   Sample set of clerical instructions

less in the sequence as presented, though it should be understood that the detailed working through of every one of these steps may be unnecessary or even infeasible in more complex real-world cases. Working through them in detail at this stage however has the virtue that it provides insight into the fairly rigorous internal structure of the decision table.

Here are the tasks in sequence:

1 *Acquire knowledge of decision logic.*
2 *Set up the table skeleton.*
3 *Identify and name the conditions and enter the stubs.*
4 *Identify and name the actions and enter the stubs.*
5 *Set out the full simple-rule specification.*
6 *Insert the appropriate action entries.*
7 *Consolidate rules where possible.*
8 *Perform rule count checking.*
9 *Append table metrics.*
10 *Arrange table layout for optimum usage.*

To illustrate the performing of these tasks we shall use the sample set of clerical instructions shown in Fig 6.5. The simplicity of this example is to aid clarity of method, though the example is by no means as trivial as it may at first seem. (It is however entirely fictional!)

Task 1   *Acquire knowledge of decision logic*
We have already assumed that this has been completed. The systems analyst, however, for the more complex descriptions may like to set out the information in some kind of chart or tree in order to get a better visual interpretation, though care must be taken not to introduce errors while doing this.

Task 2   *Set up the table skeleton*
This is just a matter of having the decision table format in front of one, either literally or in one's mind. Some organisations have in the past prepared special decision table forms as part of normal documentation and these can of course be used as worksheets. The more recent advent of simple expert system shells on microcomputers may also help with this and the interactive working with a VDU screen can be very convenient.

Task 3 *Identify and name the conditions and enter the stubs*
The task of spotting the conditions ranges from being straightforward to being fraught with problems and traps. An obvious approach is to seek out the prepositional clues such as 'if', 'for' and 'when' or adjectival phrases such as, in this example, 'born in the UK'. But one must be careful to ensure that conditions are singular (remember they were derived as Boolean variables) rather than compound, not redundant (as 'male' would be if we have already designated 'female' as a condition name), and sufficiently substantial for the context ('born' is obviously not a sensible sufficient condition to stand alone here). Bearing such pitfalls in mind it is possibly helpful to use coloured underlining or marker pen to identify candidate conditions. If we do this with our example we should arrive at something like the mark-up shown in Fig 6.6.

Thus we can draw up candidate conditions as shown in Fig 6.7, the apparent redundancies having been grouped and condensed into one stub as shown. The final condition listed is queried rather than entered as a stub

> If the applicant is female use the green form and insert '4' under 'category'. If male and a UK citizen use the green form and insert '2'. For males born in the UK but not now (for any reason) a UK citizen use the blue form and insert '1' under 'category'. For all other cases use the blue form and insert '3'.

Fig 6.6 A marked-up version of Fig 6.5

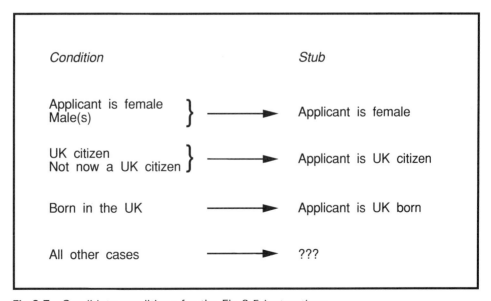

Fig 6.7 Candidate conditions for the Fig 6.5 instructions

because at this stage we cannot be sure whether it is 'other' in the sense of covering other (unknown) conditions or whether it is in the sense of other sets of condition values which have not been covered in the instructions. We shall deal with this in due course.

Task 4 *Identify and name the actions and enter the stubs*

This should be a relatively easy task, sifting out of the non-marked text what appear to be direct instructions. We have designed our example to be fairly straightforward in this respect, since it is the analysis of the decision logic rather than of the action syntax that is our chief concern. The actions are in fact of two types only (selecting the appropriate colour of form and inserting a code on the form) and should therefore all be phrased in the same way within each type.

Note however that we avoid the compound action stub, whereby the form selection and the insertion of a code would be one action stub. This enables us to analyse the actions at the simplest common level. It does however introduce a problem of sequential dependency—a code can only be inserted after the selection of the correct colour of form. For now we shall make this sequence implicit by ensuring that form selection stubs appear higher in the list than the code insertion stubs.

All this will give a candidate list of action stubs as follows:

> Select green form
> Select blue form
> Insert code 1 under Category
> Insert code 2 under Category
> Insert code 3 under Category
> Insert code 4 under Category.

Table 6.1 *Intermediate stage of development of the decision table*

CORRECT CATEGORY CODING OF FORM ACCORDING TO APPLICANT STATUS								
Rule	1	2	3	4	5	6	7	8
A: Female applicant	Y	Y	Y	Y	N	N	N	N
B: UK citizen	Y	Y	N	N	Y	Y	N	N
C: UK born	Y	N	Y	N	Y	N	Y	N
D: Select green form								
E: Select blue form								
F: Insert code 1								
G: Insert code 2								
H: Insert code 3								
J: Insert code 4								

The condition and action stubs can now be entered into a skeleton decision table format. We can also adopt a suitable (though possibly provisional) title to the table which should aim to eliminate unnecessary repetition within the stubs. This might give us then the layout of stubs shown in Table 6.1. Row labelling has been included to aid referencing.

Task 5 *Set out the full simple-rule specification*
This entails generating the appropriate Y/N pattern in the top right-hand quadrant and has already been carried out in Table 6.1. Also, rule referencing has been added.

Task 6 *Insert the appropriate action entries*
This is the 'meat' of the analysis. A blow-by-blow account would make for tedious reading and therefore we show in Table 6.2 the outcome of the work and make a few explanatory comments.

Rules 1 to 4 are easy to complete because they happen to cover the four cases of female applicant for which the action is the same whatever the UK status. This is fortuitous in a way. It has occurred because this general case was mentioned first and led us to make the female-applicant condition (row A) the first in the list where it was allocated the first block of Y entries. We also have to infer from the text that both the UK status conditions (rows B and C) have no bearing on the actions for female applicants. This cannot be proved as a logical deduction, it simply makes the most sense. If in any doubt, the systems analyst must find the appropriate authoritative source for these procedures and conduct further investigations.

The same might be said about rules 5 to 7. It is necessary to be quite clear that for a male UK citizen it does not matter whether or not he was born

Table 6.2 *The decision table fully completed*

CORRECT CATEGORY CODING OF FORM ACCORDING TO APPLICANT STATUS								
Rule	1	2	3	4	5	6	7	8
A: Female applicant	Y	Y	Y	Y	N	N	N	N
B: UK citizen	Y	Y	N	N	Y	Y	N	N
C: UK born	Y	N	Y	N	Y	N	Y	N
D: Select green form	X	X	X	X	X	X	–	–
E: Select blue form	–	–	–	–	–	–	X	X
F: Insert code 1	–	–	–	–	–	–	X	–
G: Insert code 2	–	–	–	–	X	X	–	–
H: Insert code 3	–	–	–	–	–	–	–	X
J: Insert code 4	X	X	X	X	–	–	–	–

in the UK for him to be coded 2 on the green form. Whereas with a male not born in the UK he must definitely not be a UK citizen in order to be coded 1 on the blue form.

This leaves us with one rule, rule 8, unaccounted for. The instructions say 'for all other cases', so there is a suspicion here that either our interpretation has missed some other conditions somewhere or that the original instructions have been rather loosely composed. A return to the source is most certainly advisable, but since we cannot do that here we shall assume that the plural 'cases' was wrong and that what was meant was 'for the remaining case'. It would be better anyway if the written instructions had spelt out this remaining case for the sake of clarity to the clerk, but such looseness in the use of English narrative is actually very common. The example helps to illustrate how applying the decision table analysis does help to identify such ambiguities.

Task 7 *Consolidate rules where possible*
If we look at rule columns 1 to 4 in Table 6.2, we see that all of the rules lead to the same action-pattern, which in Boolean terms means that they belong to the same Boolean function. It is also noticeable that while the first condition (row A) is always Y the values in the second and third (rows B and C) take all possible pairings—in other words their value has no bearing on the function. We could therefore express the function more simply by replacing these four rules with the single rule:

Conditions = Y   Actions = X
         —   —
         —   —
                             —
                             —
                             —
                              X

This is, in effect, K-map simplification in another guise. But it is inconvenient to resort to K-maps in this context and so we use instead the consolidation procedure explained in Section 4.3.3:

> Where a set of rules lead to the same action pattern and within that set the condition entries of one or more rows appear in all possible combinations while the remaining row condition entries remain constant, then the rule-set can be consolidated into a single rule with don't-care entries inserted in the variant rows.

The variant rows in Table 6.2 were the rows B and C, while the row A entry remained constant. Now notice that the consolidation rule can be applied also to rules 5 and 6. They lead to the same action-pattern and the row A and B entries remain constant while the row C entry takes all possible values (only two in this case, being a single binary condition). The fully

170

Table 6.3 *The decision table fully consolidated*

CORRECT CATEGORY CODING OF FORM ACCORDING TO APPLICANT STATUS				
Rule	1′	5′	7	8
A: Female applicant	Y	N	N	N
B: UK citizen	–	Y	N	N
C: UK born	–	–	Y	N
D: Select green form	X	X	–	–
E: Select blue form	–	–	X	X
F: Insert code 1	–	–	X	–
G: Insert code 2	–	X	–	–
H: Insert code 3	–	–	–	X
J: Insert code 4	X	–	–	–

consolidated table now looks like that shown in Table 6.3. The consolidated rules have been labelled 1′ and 5′ respectively. (If further consolidation of consolidated rules were to transpire this could be signalled by labelling with b, c, etc.). Rules 7 and 8 will not consolidate because they do not belong to the same function—their action patterns differ.

Task 8 *Perform rule count checking*
Having arrived at a consolidated table through several applications of the consolidation rule, the rule count check is merely a check on one's own

Table 6.4 *Rule count check on the consolidated table*

Rule	1′	5′	7	8
A: Female applicant	Y	N	N	N
B: UK citizen	–	Y	N	N
C: UK born	–	–	Y	N
D: Select green form	X	X	–	–
E: Select blue form	–	–	X	X
F: Insert code 1	–	–	X	–
G: Insert code 2	–	X	–	–
H: Insert code 3	–	–	–	X
J: Insert code 4	X	–	–	–
Rule count check	4 +	2 +	1 +	1
	= 8 = 2^3			
	CORRECT			

171

working. The number of don't-cares in the condition entries of a rule will reveal how many rules have been consolidated into it. One don't-care implies two rules consolidated, two implies four, n implies 2 power n, and so on. The total of these counts should equal 2 power m, where m is the number of condition rows. The applied check is illustrated in Table 6.4.

The check has more significance than might be apparent from such an example. It is quite likely that in a more complex example the systems analyst might condense tasks 5 to 7 into a single exercise of specifying the consolidated rules direct from the source material. In fact this could have been done quite easily in our simple example—rules 1' and 5' could both have been arrived at through a direct conversion of the narrative. Such directly specified consolidated decision tables may well then contain errors or incompletenesses since they will not have been derived via methodical setting out of the full condition-entry combinations. A failure in the rule count check will signal that something is amiss but won't of course pinpoint the fault itself.

Task 9 *Append table metrics*
There are a number of quantitative values which may be worth generating as useful descriptors of an actual decision table. The simplest of these are the three simple counts of

> number of condition rows (c)
> number of action rows (a)
> number of rule columns (r).

In our example, the final table has $c = 3$, $a = 7$ and $r = 4$. These are independent factual descriptors, but their relative values may give indications of types of table.

The value of c is the most sensitive indicator of table size and complexity since it is 2 power c that tells us the total number of simple rules implicitly covered by the table. Let us call this value R.

The actual number of rules, r, cannot exceed R, and the ratio r/R gives some indication of the extent of consolidation that the final table incorporates. Also if $a > r$ this will indicate that at least one rule must be a multi-action rule. If $a = r$, we should expect this to mean that each rule leads to one action only.

A potentially useful type of measure is that of the relative frequency f of occurrence of each rule. This can be expressed as a percentage and the sum of all f should come to 100%. Let us add some imaginary values of f to our sample table as shown in Table 6.5. These are values that can only be added when the real-world use of the table's rules has been monitored. They might be literal values if the situation is entirely stable. Much more likely is that they represent average values or perhaps even only probability values. In some circumstances it might be possible only to attach subjective values, based on best guesstimates provided by the experienced clerical workers in

Table 6.5 *Decision table with incorporated relative frequency values*

CORRECT CATEGORY CODING OF FORM ACCORDING TO APPLICANT STATUS				
Rule	1′	5′	7	8
A: Female applicant	Y	N	N	N
B: UK citizen	–	Y	N	N
C: UK born	–	–	Y	N
D: Select green form	X	X	–	–
E: Select blue form	–	–	X	X
F: Insert code 1	–	–	X	–
G: Insert code 2	–	X	–	–
H: Insert code 3	–	–	–	X
J: Insert code 4	X	–	–	–
Rel. freq. (%)	25	60	5	10

a current system. Where the table's rules are followed in, for example, different office environments then it is likely that different values will obtain in the different environments.

Task 10 *Arrange table layout for optimum usage*
In general, there are many features of layout to a decision table that are fortuitous and/or arbitrary. These include:

The wording of condition and action stubs
The sequence of condition rows
The sequence of action rows
The sequence of rule columns.

Before deciding that we have a table ready for final presentation and use therefore, we should check through each of these features and consider possible improvements. Again this work must be related to knowledge of the real-world situation.

For example, it might be discovered that users are more used to checking that applicant is male rather than female. It is the same condition, but we could change the stub for row A to read 'male applicant' and we would then have to reverse all the Y/N values along that row.

It might also be better for the user if the row C condition were read before the row B condition and that it was therefore desirable to interchange these two rows. Taking the row labels along with the rows means ending up with a non-alphabetical sequence in the labels but, until a definitely final table is arrived at, it is best not to change the labelling. In our present example we

Table 6.6 *Final arrangement of the rules*

CORRECT CATEGORY CODING OF FORM ACCORDING TO APPLICANT STATUS				
Rule	5′	1′	8	7
A: Male applicant	Y	N	Y	Y
C: UK born	–	–	N	Y
B: UK citizen	Y	–	N	N
D: Select green form	X	X	–	–
E: Select blue form	–	–	X	X
F: Insert code 1	–	–	–	X
G: Insert code 2	X	–	–	–
H: Insert code 3	–	–	X	–
J: Insert code 4	–	X	–	–
Rel. freq. (%)	60	25	10	5

shall not contemplate a change to the action rows, though there could well be a need for this in other examples.

For the rule column sequence it might be advisable to present the rules in descending value of relative frequency, since in this way the table user will encounter the earlier rule 'hit' more frequently. This is particularly important if we are looking at a potential specification of program coding sequence, where it means the shortening of the most frequently traced run paths. Of course, programmers can be instructed to code the rules with higher f values first anyway, without actually presenting them in that sequence. However, for illustrative purposes we carry out all these modifications to layout here and show the result in Table 6.6.

Note that had we not been able to obtain any f values for one reason or another the rule sequence in Table 6.5 and earlier would be 'correct' in the sense that the don't-cares appear as far to the left and as far down the rows as possible. This gives the table readability in showing the most consolidated rules first, that is to say the rules requiring fewest checks of the conditions in order to make the rule 'fire'.

6.3.2 Extending the entries

Although we have introduced the decision table format as an adaptation of the truth table the fact is that it is not usual to restrict its usage to binary logic. To illustrate this let us look at a simple *extended-entry* table with a multi-valued condition structure (see Table 6.7). It should present no

Table 6.7 *Example of a decision table with extended entries*

Interview rating Number of A levels	Excel- lent 4+	Excel- lent 3	Excel- lent 2	Good 4+	Good 3	Fair 4+	E L S E
What to do	Accept	Short- list	Reserve list	Short list	Reserve list	Reserve list	Reject

difficulty as a table to be read in common-sense fashion, even though it is indeed based upon the decision table format.

Notice the introduction of what is called an *else rule*, the function of which should be obvious. If none of the other rules is found to apply then the else rule is obeyed. Clearly there can only ever be one else rule in a table. Whilst it is a useful abbreviation for apparently all condition entry sets not otherwise appearing as rule components, it does introduce certain ambiguities that may be difficult to resolve. For example, what if there are 'hidden' interview rating entries such as 'Poor' or, even more confusing, 'Postponed interview'? It is not immediately obvious that the action should (probably) be to reject.

It is now worth taking a closer look at the effect of extending condition entries to be other than YES, NO and – for don't-care. For we have, relatively painlessly, stepped over the bridge into multi-valued logic territory!

6.3.3 Features of extended-entry decision tables

By extending entries to be able to take more than binary values we are recognising that there is often dependency between apparently separate binary conditions that is only clumsily represented by the limited-entry table. Suppose that a certain procedure requires that the condition of a used vehicle be checked as being 'as new', 'good', or 'unacceptable' as pre-conditions for some set of actions. The binary condition stubs would have to be something like:

Vehicle as new?	Y	N	N
Vehicle condition good?	N	Y	N
Vehicle unacceptable?	N	N	Y

The fact that a YES response to any one condition means an inevitable NO response to the other two is sure indication that the conditions are not independent. In a limited-entry table this means that there are a number of 'impossible' rules, where there are two or more YES entries in any one column. By combining this family of binary condition stubs into a single extended-entry condition stub, these impossible rules simply do not arise:

Condition of vehicle | new | good | reject |

We do have to generate coded mini-stubs as condition entries it is true, but, if volume of data is a problem, code-sets can be resorted to:

Condition of vehicle (new = N, good = G, reject = R)	N	G	R

which is in the same vein as using Y/N for YES/NO.

The Multiplication Rule referred to in the previous chapter will tell us that the number of possible rules in an extended-entry table is the number obtained by multiplying together all the set enumeration values for each condition row in the table. Suppose the above condition is one of three rows in some table:

Condition of vehicle
    (new = N, good = G, reject = R)
Vehicle type
    (B = bus, C = car, V = van, T = truck)
Main colour
    (code range = 01 to 12)

Then the total number of possible rules is $3 \times 4 \times 12 = 144$. This is a large number indeed and is an indication of the kind of 'combinatorial explosion' that is all too characteristic of real-world knowledge. In actual practice the large proportion of these possible rules would probably not be used since manual clerical operations in any case just cannot cope with such variety of detail. We shall simplify the example a little so as to be able to cope with the variety and better see what logical rules are in force here.

Table 6.8 shows a complete table with the same conditions but with far fewer potential rules by virtue of there being only three colours! The total number of possible rules is now just $3 \times 4 \times 3 = 36$. Each don't care must now be valued at the enumeration value for that row and the Multiplication Rule applied to any column with two or more don't-cares in it to get the rule count check values. Thus rule three represents the consolidation of $4 \times 3 = 12$ simple rules. Note also that we have used extended entries for the action row D. This has no effect on the rule count, it merely reduces the number of action rows.

But there appears to be a problem in that the check total is greatly in excess of the expected value. This brings into the limelight questions of completeness, rule overlap and consistency, which we look at in the next section.

Table 6.8 *Rule count check with extended-entry table*

PLACEMENT OF USED VEHICLES							
Rule	1	2	3	4	5	6
A: Condition of vehicle (new = N, good = G, reject = R)	N	N	G	–	–	R
B: Vehicle type (B = bus, C = car. V = van, T = truck)	C	V	–	B	T	–
C: Main colour (B = black, W = white, C = colour)	–	W	–	–	–	–
D: Display area (S = showroom, F = front, R = rear)	S	S	F	R	R	–
E: Park at industrial estate site	–	–	–	–	–	X
Rule count check total = 3 = 46		+ 1	+ 12	+ 9	+ 9	+ 12

6.3.4 Completeness and consistency analysis

If our rule count check had produced a total less than the total of possible rules (36 in our example above), we could guess that the table was incomplete in the sense of being short of some of the possible rules. There might well be a reasonable explanation for this such as when the missing rules are 'impossible' or 'never arise in practice', but at least the check can lead the analyst to explore the gaps and get users to justify their existence. But in our most recent example we appear to have more rules than all the possible ones. The explanation is that there is overlap between various pairs of rules in the table, which means that they are non-disjoint in the sense that there are one or more simple rules which have been consolidated into both.

This has happened here probably because the table has been specified direct from users' descriptions of procedures. Let us look at rules 3 and 4 for example. Rule 3 will have been derived from the user saying something like "We place all vehicles adjudged to be in good condition in the front parking area" and rule 4 from something like "All coaches (called buses here to get a distinct code from C = car) are placed in the rear parking area". So which rule applies for coaches in good condition? You may well use common sense to decide that rule 4 overrides rule 3, because you are able to call on a vast array of background knowledge and experience. (You could still be wrong though of course—there may well be room for coaches on the front parking area, in which case common sense may lead us to have doubts!)

But the decision table is only attempting to capture the bare, what we might call, logical knowledge of the rules and common sense has to be helped by direct specification if ambiguities are to be avoided. Admittedly this is pointing up how limiting bare logic is when it comes to trying to represent knowledge as it actually appears in human usage, but that is part of the purpose of the present exercise. We return to the question of getting beyond these limitations later.

What we want to do immediately is to identify and deal with all overlaps so that we can be clear about our completeness question. We can identify overlap formally (though laboriously) by checking each possible pairing of rules as being disjoint or not. In practice it is fairly easy to spot disjoint pairs quickly by noting that there are different entry conditions in a particular row. For example, rows 1 and 2 have different row B entries and cannot therefore be overlapping. In fact the identification of overlap is achieved formally by detecting that there is no such difference in any of the rows for any rule-pair.

Working methodically left-to-right we can observe that rule 1 does not overlap with any other rule in the table and the same is true for rule 2 for all rules to its right. (We do not need to check to the left again.) Rule 3 however does overlap with both rules 4 and 5; rule 4 overlaps with rule 6; and rule 5 also overlaps with rule 6. Rule 6 is the rightmost rule, so our checking is now complete.

The question then arises as to what precisely are the simple rules involved in the overlaps. These can be identified by making intersections between all row-entry pairs for each rule-pair. Thus, in Table 6.8, between rule 3 and rule 4 the overlap is

Rule 3	*Rule 4*	*Overlap*
G	& {N, G, R}	= G
{B, C, V, T}	& B	= B
{B, W, C}	& {B, W, C}	= {B, W, C}

Don't-cares have been set out in full for clarity. So we see that there are three simple rules which are consolidated within rules 3 and 4. As we had already discussed, these are the ones relating to good-condition buses (of any colour). Rules 3 and 4 lead to different actions so there is an inconsistency to be resolved. Let us assume that it is indeed rule 4 that applies in these cases, ie. all buses go to the rear parking area. We then have to remove the overlap rules form rule 3 and we can do this by making a DIFFERENCE operation in the second row entries, the only entries which do in fact differ between rule 3 and the overlap rules:

{B, C, V, T} DIFFERENCE {B} = {C, V, T}

This gives a new rule, 3′, with which to replace rule 3. We have had to admit here a new type of condition entry—a set of entry values, in this case {C, V, T}. This is to be interpreted as meaning that a simple rule is to be generated for each element of the set, so it is merely an extension of

Table 6.9 *Recalculated rule count check*

PLACEMENT OF USED VEHICLES						
Rule	1	2	3'	4'	5'	6
A: Condition of vehicle	N	N	G	(N,G)	(N,G)	R
B: Vehicle type	C	V	(C,V,T)	B	T	–
C: Main colour	–	W	–	–	–	–
D: Display area	S	S	F	R	R	–
E: Park at industrial estate site	–	–	–	–	–	X
Rule count check total = 3 = 37		+ 1	+ 9	+ 6	+ 6	+ 12

the don't-care principle to incomplete sets of condition values. The Multiplication Rule will apply in the same way when it comes to rule count checking of course.

For the remainder of the overlap cases, let us assume that rule 6 was intended to cover all 'reject' condition vehicles, so that the overlap rules have to be removed from rules 4 and 5. (Note that this means that rule 4 did not after all really mean *all* buses!) The outcome of all this is shown in Table 6.9.

Our rule count check still fails because rules 3' and 5' still overlap. (We didn't get around to removing the overlap between rules 3 and 5 before they

Table 6.10 *Further rule count check recalculation*

PLACEMENT OF USED VEHICLES						
Rule	1	2	3"	4'	5'	6
A: Condition of vehicle	N	N	G	(N,G)	(N,G)	R
B: Vehicle type	C	V	(C,V)	B	T	–
C: Main colour	–	W	–	–	–	–
D: Display area	S	S	F	R	R	–
E: Park at industrial estate site	–	–	–	–	–	X
Rule count check total = 3 = 34		+ 1	+ 6	+ 6	+ 6	+ 12

Table 6.11 *Correct rule count check*

PLACEMENT OF USED VEHICLES								
Rule	1	2	3″	4′	5′	6	7
A: Condition of vehicle	N	N	G	(N,G)	(N,G)	R	N
B: Vehicle type	C	V	(C,V)	B	T	–	V
C: Main colour	–	W	–	–	–	–	(B,C)
D: Display area	S	S	F	R	R	–	–
E: Park at industrial estate site	–	–	–	–	–	X	–
F: Refer to office	–	–	–	–	–	–	X

Rule count check total = 3   + 1   + 6   + 6   + 6   + 12   + 2
= 36 correct

became 3′ and 5′.) We now have to decide whether trucks ('new' and 'good') should have action row D or E. We shall assume that, on checking the facts, we find that no trucks are ever actually kept at the front area and therefore rule 5′ wins over 3′. Identifying and removing the overlap from 3′ to give 3″, we now have as shown in Table 6.10.

The rule count check now falls below the expected value of 36! Since there is now no overlap in the table (though this should still be checked for one last time), we can deduce that there are two rules missing. Finding missing rules can be a tiresome manual task since every possible combination of condition-entry values must be checked off against the table. This would in due course reveal that the two missing rules are:

$$\begin{array}{|c|c|} \text{N} & \text{N} \\ \text{V} & \text{V} \\ \text{B} & \text{C} \end{array}$$

The black vans and coloured vans have not been given a home. What the answer to our query about this would be we can only guess.

"We never acquire anything other than white vans" perhaps. Although we would almost certainly want to go into the reasons for this idiosyncratic policy in the real situation, we shall merely accept it here and for completeness sake enter the new rule and its associated action row F into our table (Table 6.11). The decision table would now be ready for coding into a knowledge system.

6.3.5 Rules as discrete entities

The decision table view inevitably leads us to look at rules as part of a wider system which comprises a certain number of conditions and actions. But when we design, invent, specify and use rules in human practice we tend not to see this wider system, or even to worry about it very much. If we want to see how one rule compares with another rule, in terms of shared or discrete criteria say, we are more inclined to sketch a tree-like diagram (see Section 6.4).

A rule IF {criteria to be met} THEN DO {actions}, viewed in isolation, is simply an event waiting to be triggered. The fact that it is one column among many in some possible tabular layout is almost irrelevant to this perspective. The knowledge processing terminology for this is that a knowledge system comprises a collection of **production rules** waiting to be fired by the appropriate data events. The criteria may have been met because of the firing of some other rule, and the firing of this rule may well trigger off the firing of further rules in a sequence of automatic reasoning.

This is indeed a very different perspective from that of our meticulous setting-out of the decision table. The number of criteria needing to be satisfied before a particular production rule is fired may be so large that a decision table is a massive model to contemplate and certainly not a feasible manual analysis tool. But knowing the theory and principles of the decision table model does give an insight into the inter-rule structure that is potentially always there, and knowledge systems designers may find it invaluable to fall back on this when looking at the fine detail of rule-based systems design.

The extended-entry decision table is also a useful lead into the question of multi-valued logic. The fact that the entries are no longer in binary YES/NO form does not seem particluarly shocking or too much of an intellectual leap! We shall find this useful when we come to talk of other non-binary logic and knowledge representation techniques in Chapter 10.

6.4 Decision trees in use

6.4.1 Introduction

The tree-structure is of fundamental importance in any approach to formalising intelligent processes. It is a universal model of certain kinds of organised human thinking, which very often treats problem solving as using simpler and more handlable concepts as testbeds for more difficult and even intractable problems. The following are typical:

- A host of facts to make sense of? Then generate broad categories under which to place them and if the categories are still too populous create sub-categories within them and so on.
- Too many apparent alternative ways for tackling a problem? Again let's see if we can divide the choice into a few main families and these families into sub-families, and so on.

● Too much going on for one person to control? Create sections or departments and if necessary sub-sections and so on, so that each controller only has a few sub-controllers to control.

'Controllers' don't need to be people, either. They can be budgetary or cost-centre mechanisms, program modules, screen-based menus, system design elements. Indeed, this last type is what structured systems analysis is very largely about.

True, there are other important models of effective human thinking, such as for example those based on intuition, creativity, and experience. An over-structured approach to some problems can delay a solution being found for so long that for all practical purposes the problem becomes unsolvable by that route. Or the structure itself, if made tangible, may impede the useful action, such as might happen if a menu system has too many levels, impeding the work-rate of a practised user or of one who did not need all the options in the first place!

With these reservations firmly in mind we should nevertheless seek to understand what it is about the tree structure that makes it so potentially useful as a thinking model. This seems the appropriate point to attempt this, since every junction on any tree structure can potentially be viewed as a decision point, a point at which a choice is to be made.

6.4.2 AND trees for analysis of sub-classes

The company organisation chart is the most commonly encountered AND tree in everyday use. As its name implies, all branches from a junction or node must be taken in conjunction. Put another way, every node joined directly to a higher-level node has to be taken into account before the view of that node is complete. In an organisation chart this is evidenced by the fact that it is intended to show all sub-managers under any given manager, though they might be called something different (eg. Directors, Section Leaders) at other levels.

The AND tree is also very commonly the basis for other charts to be found in business and management. Management by Objectives (MBO) sets out to establish an AND tree of objectives for the environment to which it is being applied—a chief aim is divided into sub-objectives and each of these into sub-sub-objectives and so on. Problem analysis tackled top-down usually generates an AND tree where the nodes are problem identification statements. A very general problem statement is thus progressively analysed down into ever more detail, until at the leaves (at the bottom of an inverted tree, of course) appear problem statements detailed enough to be tackled with a known amount and quality of resource.

AND trees then are very easy to follow and there is often no need to state explicitly that they are AND trees. Care should be taken however in systems analysis documentation, where trees of various kinds may be found, to make clear at all times that a tree chart is in fact an AND tree even though in its own narrow context the fact may be obvious. We return to this point in the next section also.

6.4.3 OR trees for analysis of alternatives

Charts based on the OR tree make their appearance whenever 'cascades' of options have to be considered for any reason. Here the understood reading of the chart is that at each node one only of the branching paths can be selected—that each branch represents an alternative way forward through the tree structure. Management scientists usually mean this when they use the term 'decision tree'.

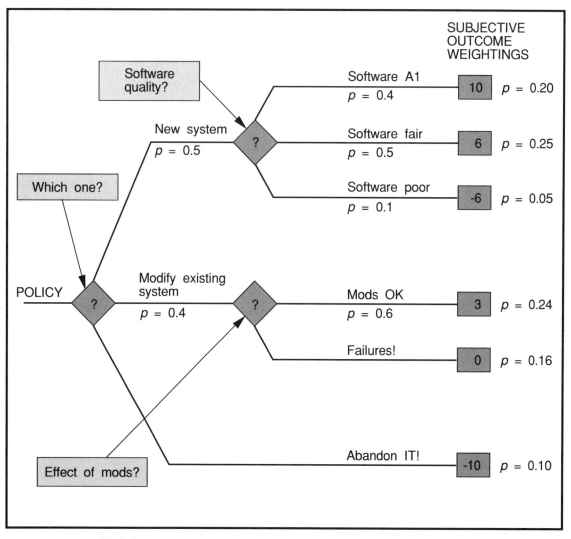

Fig 6.8 Example of a probability analysis decision tree

Such a decision tree aims to help the scientist to quantify probabilities and risks in the type of decision-making environment where different sequences of decisions can lead to (perhaps very) different outcomes. One such system of analysis is illustrated in Fig 6.8. Each branch from a node is assigned a probability, which might here represent at the first node (on the left) an estimation of the likelihood of current management acceptance, and at the two subsequent nodes the probabilities of the outcomes shown. The figures at the leaves (rightmost in this horizontal tree) might be estimates of the relative penalties/rewards of arriving at the final situation. Probabilities are accumulated (multiplied) along each path so that each leaf also has a final probability associated with it. How the probabilities and penalty/reward pairings are interpreted is a matter for any given situation. In any case it is advisable that the method be used only by an experienced and qualified management scientist. We have shown it here in simplistic form to illustrate the use of an OR tree structure.

The fact that the sum of the probabilities of branches emanating from any one node must come to 1 (unity) is related to probability theory, but it also points to an interesting feature of the tree. If seen from the point of view of the scientist creating the tree it is in fact an AND tree, because the analysis is incomplete at any node until all possible branches have been identified. So as an analysis tool it is an AND tree—only when computations are being carried out does it become an OR tree! The scientist will be aware of this switch, but it does help to remind us again that clarity of interpretation of a tree is crucially important, and that the appearance of such trees within a system's documentation should be accompanied by notes providing such clarification. Some very strange misinterpretations are possible if this is not done.

6.4.4   AND-OR trees for general analysis

AND-OR trees were introduced in Chapter 4 as a method of representing Boolean expressions. This suggests that, in general, the tree can represent all logic structures and this is confirmed by the fact that various types of AND-OR trees are increasingly to be found as part of the tool-kits for systems analysis. The 'types' here refers to types of chart convention, for there are already a number of these in evidence.

The simple device of using the arc to denote an OR between branches and the absence of the arc to indicate AND (already discussed in Chapter 4) may be sufficient for a broad-brush treatment of analysis (see Fig 6.9). It can be useful to be able to resort to the algebraic interpretation of such charts in order to clarify what the distinct alternatives actually are. They are of course the clauses of the normalised expression. For the example in Fig 6.9 these are

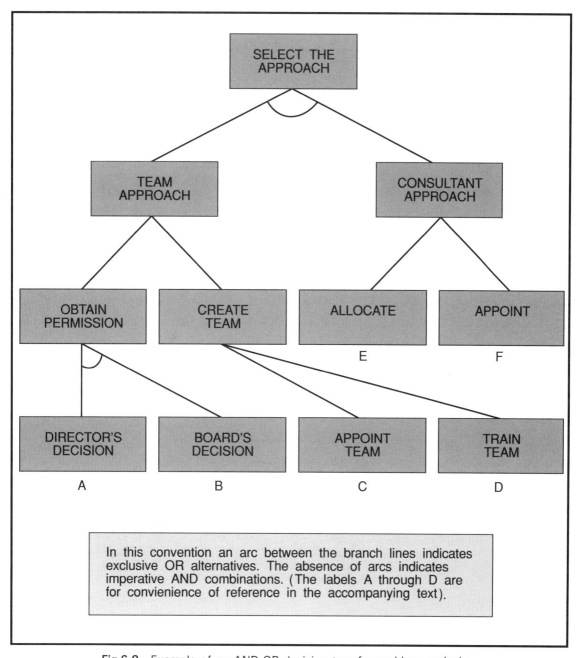

In this convention an arc between the branch lines indicates exclusive OR alternatives. The absence of arcs indicates imperative AND combinations. (The labels A through D are for convienience of reference in the accompanying text).

Fig 6.9 Example of an AND-OR decision tree for problem analysis

$$((A\ OR\ B)\ AND\ C\ AND\ D)\ OR\ (E\ AND\ F))$$
$$= (A\ AND\ C\ AND\ D)\ OR\ (B\ AND\ C\ AND\ D)\ OR\ (E\ AND\ F)$$

the alternative combinations being those within the brackets. Using these it is then possible to read out what these alternatives are from the chart.

For more precise analysis of the logic structure it is necessary to introduce

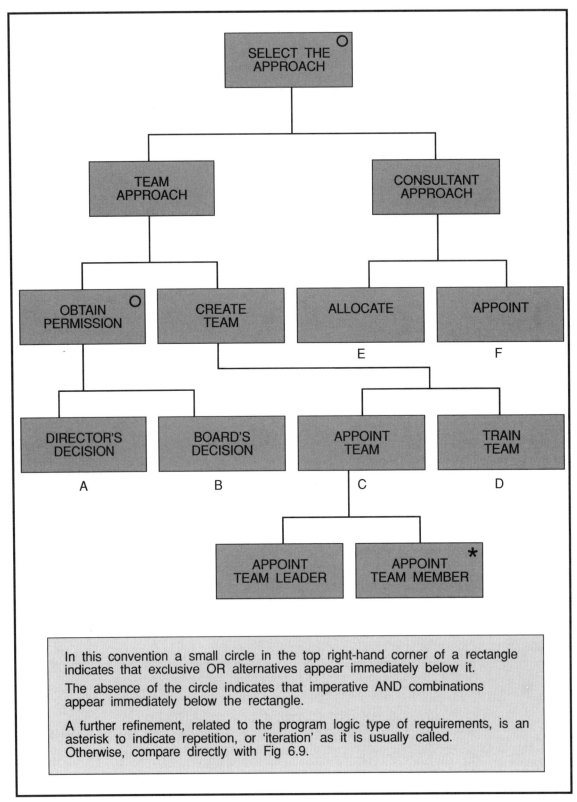

In this convention a small circle in the top right-hand corner of a rectangle indicates that exclusive OR alternatives appear immediately below it.

The absence of the circle indicates that imperative AND combinations appear immediately below the rectangle.

A further refinement, related to the program logic type of requirements, is an asterisk to indicate repetition, or 'iteration' as it is usually called. Otherwise, compare directly with Fig 6.9.

Fig 6.10 A decision tree expressed in Jackson Structured Programming (JSD) conventions

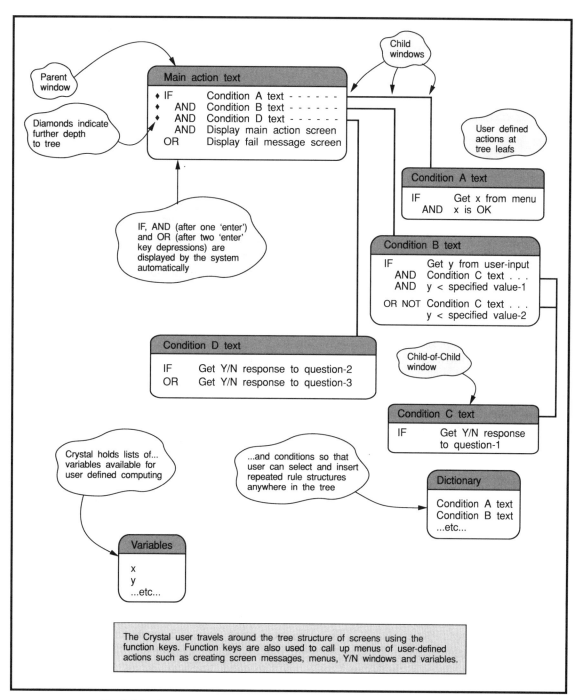

Fig 6.11 Building AND-OR decision tree structures using Crystal

further conventions. This is especially true of analysis of program structures, where iteration has to be given distinctive recognition and where it is convenient to indicate selection of an alternative from more than the binary option. Figure 6.10 shows a relatively simple example from the Jackson Structured Programming convention type. Our aim here is not to review such conventions, far less to instruct in their use, but to make the analyst aware of the fact that all of these special tree-charting conventions have a basis in AND-OR tree structuring.

Finally, it is worth noting at this stage that many expert system packaged software systems (shells) are based on AND-OR tree building facilities. Crystal, for example, offers the user a screen window into which text statements can be entered and joined by AND and OR connectives according to whether a single or double ENTER key depression is made after each statement. (How's that for being cryptic!) There are other commands for introducing variables and negation into the statement. The logical expressions thus built up are contained within an IF...THEN envelope to which the user adds functions using other windows and commands. The user builds up a knowledge-base pragmatically and interactively (rather than as a graphically displayed tree chart) and this is guaranteed to possess a clearly defined structure which the package can use to answer queries, and to explain its reasoning (which tree paths it followed). Figure 6.11 gives a schematic illustration of this.

6.4.5 A comment on search trees

Search trees have a special importance for computer science, an importance that is not only long-established but one that grows with the continuing development of artificial intelligence in applied computing. They are important in the main because the tree structure is used as the map type for planning and controlling the routes followed by computer programs engaged in seeking out a solution to problems which offer a bewildering complexity of alternative routes. In non-technical terms this is due to the need for a structure of sub-programs where the ethos of "that attempt didn't work, so what next?" applies.

The "what next?" implicitly carries with it the need to be sure that we have completely recovered from the failed attempt, that we are truly starting again with all conditions restored to the states they were in when we started the failed attempt. The tree structure, where moving onto any branch guarantees an isolated section free from connection with the other parts of the tree, is clearly well suited to the need. Such processes are termed **search processes** and therefore the tree structures which guide them are called **search trees**.

The need for search processes arises in almost any software-driven system where large amounts of data are to be frequently accessed with constantly varying data-requirement criteria. In these days of sophistication of data on

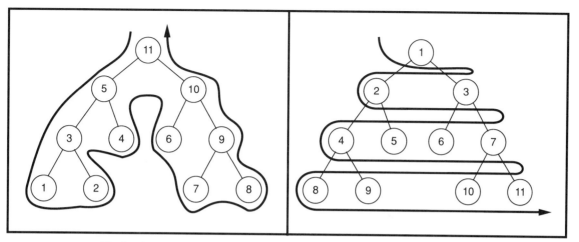

Fig 6.12   Depth-first and breadth-first tree searching

everything from the home games computer right up to the vast weather forecasting machines, this means an almost universal need.

But it is not only the volume of data to be searched that creates the demand for search process structure. In a way that is the simplest aspect of the search problem because data usually implies stored binary images of known and relatively unchanging structure. "What to do next" can be specified with some reasonable confidence by incorporating algorithms which work through the tree structure according to some fixed sequence of attempts, such as "Try the leftmost path of branches first and exhaust all alternatives from the bottom up" (depth-first) or "Try the top level first, then the second level down, and so on" (breadth-first). See Fig 6.12.

These, and variations based on them, will be familiar enough to students of computer science. A particular algorithm is selected to suit the job in hand and thus built into the software accordingly. It is not intended that we should discuss how these algorithms are designed and selected here. Proper and thorough treatment can be found in any number of computer science texts.

We will, however, briefly consider the situation where the problem resides more in the volume and complexity of selecting action to suit the nature of the problem itself. The question is then more like: "How do I decide what can be tried next and then choose the best option?" This brings the program designer face-to-face with human thought processes, which somehow know how to make judgements about appropriate action based upon interpretations of what they perceive. That perception may lead to a kind of pattern recognition, whereby the problem is recognised as being like, or even only vaguely similar to, a previously experienced problem situation. Sometimes the situation appears to be entirely new to the human's experience and yet intuition and guesswork may be brought into play, rather than bringing all thought process to a crunching halt with a message of "Will Not Compute!"

Computer scientists distinguish this latter type of situation by calling the methods for tackling it *heuristic search*. (The automatic tree searches of the depth-first, etc. type are called *blind-search* algorithms.) 'Heuristic' is more a term of negative classification, i.e. it is *not* blind-search. More positive ways of explaining it do exist such as saying that it attempts to incorporate human 'rules of thumb' into the search process. These rules of thumb depend upon human experience and may be related to recognition of what the problem is like or alternatively they may be to do with the choice of action. The choice of action itself may be based on the doubtful logic of justifications like "it seems to work (though I haven't a clue why)" or, at the other extreme, on respectable and impeccable statistical computation of which action is most likely to work based on empirical evidence.

Clearly, the incorporation of heuristics into tree structures implies that the nodes represent considerable computation centres in their own right. Not only must there be programs which are capable of making the 'which branch' decision, but these programs must have access to data files which hold the 'memory' of the experience of past decision making. This brings us to the brink of true *learning (by experience)* in artificial intelligence. The updating of the experience data in the files will mean that the branch chosen today may well not be the branch chosen at some future date. And if a program can learn to modify its decision-making behaviour at a node, why can't there be a higher-level control program which learns to modify the tree structure itself?

That takes us into a higher order of intelligent learning behaviour, where the simple tree may no longer suffice as a process control structure and where the decision making has been promoted up the scale to being rule-based reasoning. However, as we shall see in later chapters, rule-based processing itself comes face to face with this choice between blind and heuristic searching.

7 Logical Reasoning

7.1 Introduction

7.1.1 Information processing and reasoning

Information processing, served by the increasingly sophisticated facilities of information technology, has the ultimate aim that its human user should be able to base action upon a trust that what the system reports is indeed the case. If what it reports is in the form of bald factual data, then the user should be aware of carrying a considerable amount of responsibility for the transformation of the data into action. If however the system in effect reports that, from the data it has access to and using the available software which can process that data, it proposes a certain course of action, then the user has to trust not just factual data but also the correctness of the 'intelligent' software—not just the correct running of the software but also the correctness of design and the implementation of that design as actual coding.

Because no system, especially perhaps the human system as design agent, is infallible, the user has to exercise some judgement about whether what the system imparts can be relied upon. With simple data retrieval erroneous output is not necessarily difficult to detect, especially if it persists (though this is not to underplay the difficulty of de-bugging and program maintenance.) The action-advice output however is very much more difficult to evaluate for error and is much more like having to decide whether the advice of a second person can be trusted. If that second person is an acknowledged expert it is easier to trust the advice, but the social systems which support the 'acknowledged expert' credential are numerous and complex in the extreme.

Forgetting for the moment the subtler human connotations, we are talking here about trust—trust in the reasoning processes inherent in the design of the system. In order that software can be designed and written so as to be able to reason we shall have to find ways of logically modelling reasoning processes first. Indeed we ought, before that, to be sure that we ourselves have reliable reasoning techniques upon which to base the models! models!

In this chapter we shall from the outset be assuming that the last of these has been achieved. In other words we want to concentrate here upon the business of modelling the logical reasoning processes that have for the most part been accepted as valid for centuries. We need philosophers and logicians to keep challenging these reasoning processes it is true, but a glance at any textbook adopting a rigorous approach to the bases of logical reasoning will quickly show why we would be distracted from our main aims if we were to try to do the same here.

It may already have occurred to the reader that *any* piece of implemented software is a 'model of a reasoning process'. This is true. But program A may be a model of A's reasoning process and program B that of B's and so on. There is never a guarantee in the greater proportion of existing software based upon second and third generation languages that the same reasoning processes have been used from one program to the next. If things go wrong there is nothing for it but to carry out surgery on the individual program in order to track down the error in the reasoning there. What is needed is a common concept of the reasoning process so that all programs follow the same reasoning pattern. That is an ideal which can never ultimately be achieved for a number of reasons but it is a good target upon which to set our eyes.

7.1.2 Data-retrieval reasoning

Let us look first of all at a relatively straightforward information process and examine it for its reasoning content. Every process starts with some *given facts*, some *rules* which can be worked through, some *means of working* the rules, and, to trigger the process, some kind of *goal*, and finally an *outcome* to the triggered process. A data-retrieval process could then be seen to comprise, say:

> *Given*: a file or database.
> *Rules*: an enquiry handling program.
> *Means*: an appropriate computer.
> *Goal*: request for data to be output.
> *Outcome*: output of data

Let us define this process more precisely to keep things simple. We are concerned, shall we say, with getting a response to the input A?, which stands for the question: Is fact A in the database? We need to be more precise now in looking at the process components (although we shall quietly drop the *means* as being understood):

> *Given*: a database containing data A.
> *Rules*: a program incorporating the rule: If input is A? and A matches with an item in the database, output Yes.
> *Goal*: respond to A?
> *Outcome*: output Yes.

Looked at in this way, we can recognise that the rules carry rudimentary reasoning skills. This reasoning process is triggered by being set an objective (the *Goal*) which it must be able to interpret, and is then able to appropriately refer to its own store of memory (the *Given*) so as to complete the task with an appropriate response (the *Outcome*).

How general is this reasoning ability? If we assume that like the data A there could be some other data B which is also in the database, it would be disappointing if the program were not able to cope with the query B? using the same portion of coded instructions. The only difference would be that the data B would need feeding into the code in the right way. The rule might then be viewed in a more general format as being:

> *Rules*: A program incorporating the rule: If input is x?
> and x matches with an item in the database, output
> Yes.

where x stands for every recognisable item of data in the database. We begin to sense perhaps that the rule involving x is independent of A, B, etc., but that the coding which feeds A and B to that rule is not—at least in the sense that it has to be able to distinguish between them! We could then think of the rule as being a *Given* and the coding that feeds it as doing the work. This initiates the concept of a **rulebase**, which is a memory store of rules in the same way as the database is the memory store of data.

But we might spot that this 'rule' is not quite as general as we thought, for isn't it actually itself a special case of the structure IF...AND...THEN...? It is! So the rule itself could be generated by storing the IF...AND...THEN... rule in the rulebase and ensuring that there is coding capable of turning this rule into the required specific one. But now we realise that IF...AND...THEN... is actually a special case of IF...THEN..., so let's store the latter in the rulebase and trigger it as necessary.

We have gone far enough with this apparently very simple example, that of querying a database to discover whether there is certain data present or not. For what we wanted to show is that rules can be treated as processable data and, further, that rules can be treated as processable rules. How generalised the rules in any particular rulebase may be depends upon a number of practical issues. For example, a software system that aims to 'watch' an image of a speaker and 'lip-read' the queries being put to it will need to have resort to very fundamental and general rulebases to cope with the enormous potential variation of input, whereas an expert system shell which allows only fairly specific input for processing as logic will only need to have a rulebase specific to its own defined interactive syntax. (If it is much more general than it need be, it will be inefficient because of having to go back to basic principles continually.)

To distinguish between the intelligence contained within the rulebase and that inherent in the interpretive and triggering software around it, the latter is often referred to as the **inference engine**. Since there is, in principle, little reason to separate data and rules, the collective base is referred to as a **knowledge base**. In practice both inference engine and knowledge base are in any case likely to refer to groupings of software with more specialised identifiable functions. We shall return to this discussion later.

7.1.3 Reasoning with propositional logic

If we are to think about building a bridge from what we have established as rules of logic across into the issues of the generalised rule building mentioned in the previous section, it is natural to turn to the interpretation of two-valued logic as the processing of propositional statements which we first introduced in Chapter 2. For one thing, the statements are most like the phrasing which appears in the rules and, for another, the two-valued system gives the simplest and clearest representation of truth valuing. Though much more than this is demanded of representation systems by real-life knowledge processing as we shall be discussing in the final chapters, it is better to base more sophisticated systems on a simpler 'starter' system that works and then build (or discard) as necessary as progress is made.

First we have to accept that our statements must conform to the two-value requirement. This is a very constricting requirement that we will seek to loosen as soon as possible, but it must be our starting point. Thus, while we might have a few reservations about accepting

The applicant is male

as a propositional statement, it would be far less comfortable to adopt

The applicant is good

unless perhaps 'good' is a grading box on an assessment form. If the second of the above is accepted it must be on the understanding that it is computable as TRUE or FALSE within the sphere of application of the logic system of which it is a part. This will almost certainly involve more careful expression of the statement in practice, such as (assuming the assessment-form meaning applies)

The applicant is graded 'good' on the assessment form

There may be other ways of expressing this same point, which reminds us that the statement is intended to represent an intrinsic truth about the world which can actually be represented in (probably) an infinity of ways. This is familiar enough in programming, for example, where the name of a variable may be whatever the programmer likes (though the choice is unlikely to be literally infinite in practice).

In addition to potential propositional statements of the two-valued type, we have of course all the rules of the two-valued logic to call upon. So our database is represented by appropriate propositional statements with assigned truth values; our rule base may comprise whichever of the rules of Boolean logic we expect to be useful; and our inference engine by working through with logical rigour to 'prove' the answer to whatever query arises using the given knowledge base. The remainder of the chapter is addressed to this challenge.

7.2 Inference and argument

7.2.1 The material versus the logical

We have already met a number of formats for logical inference as a connective between two Boolean variables. We can list the most common formats as

A IMPLIES B
$A \rightarrow B$
$B \leftarrow A$
IF A THEN B
B IF A

All these express the same relationship between A and B, a relationship which is unambiguously defined by means of the truth table shown in Table 7.1 (using T = TRUE and F = FALSE, rather than 1 and 0, to better match our present context).

If we are to be certain that this truth table does apply between the two variables A and B, we would need to know about the actual meanings of A and B and believe what we learnt. For example, suppose that A and B represent the propositional statements:

A: It is raining
B: The roof is wet

Can we believe that the table does apply to this pair? There is a great deal more information that we would need to be given in order to be convinced. We might ask such questions as

Table 7.1 *The truth table for* A IMPLIES B

A	B	$A \rightarrow B$
T	T	T
T	F	F
F	T	T
F	F	T

- Does A mean that it is raining hard enough and has been raining long enough to guarantee the wetting of ground areas beneath the rain cloud?
- Is the roof beneath that rain cloud?
- Is the roof free from shelter from the rain both from above and from the side?
- Is the roof unheated?

If all these questions are answered to our satisfaction in the affirmative, we might be ready to accept that A will cause B to be true. But for the table to apply we also have to be satisfied that B can result from other causes. Thus we might ask

Can the roof be soaked by other means, such as by hose-pipe?

If this is shown to be the case we might finally agree to go ahead with allowing the relationship between A and B to be represented by Table 7.1. Admittedly, more meticulous investigation might be necessary according to context, such as defining wetness, extent of roof, level of precipitation and so on. The point is that a sufficient knowledge about the real-world meanings is necessary in order to accept that the implication connective does apply here.

It is therefore usual to describe such modelling as material implication. This is partly to distinguish it from the 'purer' implication of logical deduction. Let us set out an example of logical deduction using our rain example:

If it rains then the roof gets wet.   (Material implication)
It is raining.   (Material fact)
I deduce that the roof is wet.   (Logical deduction)

If we look at an algebraic representation of the above it might look like:

If it rains then the roof gets wet. $(A \rightarrow B)$
It is raining. $(\& A)$
I deduce that the roof is wet. $(\rightarrow B)$

or as a single expression:

$$((A \rightarrow B) \& A) \rightarrow B$$

which, being a Boolean expression, we can fully evaluate. The evaluation is shown in Table 7.2.

The result is tautologous—it is true whatever the values of A and B. The implication sign preceding B in the final column is logically valid. But it is a true representation of a relationship *due to the inevitability of the logical rules* rather than due to our knowledge of the real world. This is *logical implication* as distinct from *material implication*. The step-by-step use of rules to draw certain conclusions is called the *syntactic turnstile* by logicians;

each application of the rules (the syntax of the language system) is a one-way click of the logical gate. It is one-way because implication is one-way.

This difference between material and logical implication can be puzzling to the non-mathematical inquirer. To try to further clarify that there is a difference, let us look at a truth table involving only the participants in the logical implication result by masking out the intermediate columns from Table 7.2. To emphasise the antecedent role of the expression we shall represent $(A \rightarrow B) \& A$ by the single symbol C (see Table 7.3).

This is curious. The first column does not contain the 'correct' pattern of T/F values, giving rise to a redundancy at the second and fourth rows. Let us remove the redundant second row and replace it with the 'correct' values (see Table 7.4). The truth table is incomplete because we are unable to insert

Table 7.2 *Evaluation of the expression* $((A \rightarrow B) \& A) \rightarrow B$

A	B	$A \rightarrow B$	$(A \rightarrow B) \& A$	$((A \rightarrow B) \& A) \rightarrow B$
T	T	T	T	T
T	F	F	F	T
F	T	T	F	T
F	F	T	F	T

Table 7.3 *A partial view of Table 7.2*

$(A \rightarrow B) \& A$ [ie. C]	B	$C \rightarrow B$
T	T	T
F	F	T
F	T	T
F	F	T

Table 7.4 *The amended Table 7.3*

$(A \rightarrow B) \& A$ [ie. C]	B	$C \rightarrow B$
T	T	T
T	F	(not specified)
F	T	T
F	F	T

a truth value in the result column at the second row. The reason for this is that it is not possible for C to be TRUE unless both A and B are TRUE (see Table 7.2) *because of the meaning we have ascribed to C*, that is to say its being equivalent to (A → B) & A. There has already been one click of the syntactic turnstile.

Although we can think of the steps of an argument as being clicks of a logical gate that cannot be turned back, this doesn't prevent keeping a record of what steps *have* been taken so as to be able to scan it in reverse direction. We are able to 'tell the story' of an argument if need be. But following the story of an argument is a different mode of reasoning to working the argument through in the first place. An expert system will use its inference engine software to process a request for knowledge, but it requires a different type of 'tracing' software to be able to explain afterwards to the user how it arrived at the result it did.

This modest enquiry has shown that it is vitally important to be clear about the starting knowledge (what is given) in logical reasoning. Such data is referred to as the *premises of the argument*. Not being clear about the premises is like not knowing which database is loaded when launching a query against it. Furthermore, not being clear about the pre-existence of a click of the semantic turnstile among a set of apparent premises (they are not strictly premises if such is the case) is rather like starting up a program without all the parameter values being properly initialised.

The mathematical logician has a pair of symbols to make clear certain kinds of truth when setting out logical argument. These are

$\vdash E$ It is asserted that E is true
$\vDash E$ It follows that E is always true

Note that both can be read as It is true that E is true; but the first is concerned with assertion (before proof or demonstration) and the second with tautology. They are concerned with describing in shorthand the state of the argument rather than being part of the logical symbolism.

Premises can be asserted:

$\vdash A$
$\vdash A → B$

Or we can assert that it is going to be proved that something is true:

$(A → B) \& A \vdash B$

Having proved it we can then write:

$(A → B) \& A \vDash B$

and finally seal this as a rule that is universally true and usable in subsequent argument by writing:

$\vDash ((A → B) \& A) → B$

We see that the symbols are used relative to context. If we amplify the second symbol to mean

$\models$ It is therefore tautologous at this stage of the argument that ...

then at the appropriate stage in our earlier enquiry it could have been stated that

$$\models C \to B$$

even though it looks, out of context, to be a statement of material implication. Taking argument to read line-by-line downwards (we discuss presentation of argument in a later section), a very simple example of this is the argument:

$$\vdash A$$
$$\models A$$

This states that if it is given that A is TRUE then it is tautologous that A is TRUE! In algebraic terms this is to say that the formula $A = A$ is tautologous (as also, but separately, is the formula $\tilde{A} = \tilde{A}$). Although this may seem obvious, it is fundamentally important to any formal knowledge processing system to say that what is true is indeed true. Imagination and dreams may be able to cope with failures in this basic argument, but formal systems cannot—though it might be worth again giving a little thought to the importance of time in this matter. A is true today, but may not be tomorrow, in which case $A = A$ no longer holds! We return to these issues in the final chapter.

Another little puzzle in data and knowledge processing arises when we know that a fact may exist ('applicant is either male or female') but do not yet have the value assigned. It can be important not to treat both 'applicant is male' and 'applicant is female' as FALSE even though their opposites are not TRUE. Whether this is treated as a time problem again, whether a different category of logical value other than TRUE and FALSE is called for, or whether we make an assumption of TRUE or FALSE and then proceed as normal, will depend upon the logic system adopted by the database system. Sometimes this is dangerously undefined or incompletely defined and bugs may proliferate as a consequence.

7.2.2 The general process of argument

Argument is a general term applied to the proper following through of logical reasoning. We have been using rather loose argument throughout this text in the interest of readability. But we have seen in the previous section that we do need to be careful about starting points, and the need for following recognisable rules and conventions when working through reasoning processes hardly needs emphasising. Mathematicians have learned that they need to work back to the extremes of abstraction in order to set up truly reliable *formal systems*. There is the fundamental problem that in order to state the basic rules of such a system, the rules of the rule-description language must be defined, thus creating a recursive problem of never being able to arrive at the set of rules which represent the innermost Russian Doll.

In general, the answer has been to stop recursion at the level of abstraction of a set of symbols (any will do!), write down strings of these symbols which are defined as being in acceptable sequences, thus arriving at what, it should be remembered, are then termed *well-formed-formulae* (wff). Decimal arithmetic thus comprises the symbols 0 through 9, four operating symbols for addition, subtraction, division and multiplication, a pair of bracket symbols and an equals sign. The string

$$(1 + 2)/3 = 1$$

would be a wff in this system.

The true starting point must be a set of wff which can be written down without reference to anything else—these are called the *axioms* of the system. Then a set of rules which enable further wff to be generated from the axioms are posited—axiomatic *rules of inference*. A sequence of rule applications may then give rise to further, more remote, wff and these stored away as *theorems* for future reference. A theorem may be put up as requiring proof, the proving exercise being the process of using axioms,

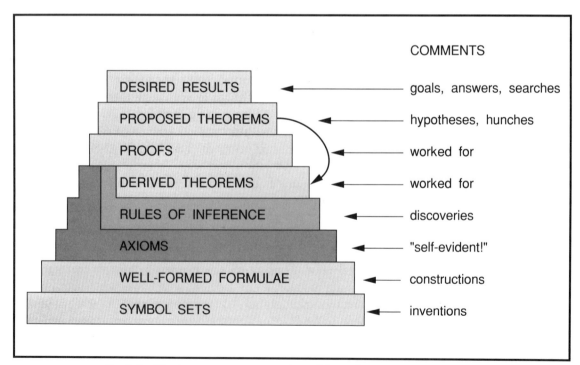

Fig 7.1 The bottom-up construction of formal systems

rules and previously accepted theorems to arrive at the wff proposed by the theorem. The structure of this system is outlined in Fig 7.1. (Warning: even this description would fall foul of the pure formalist, because it does not distinguish between true theorems based only on axioms and derived theorems based on sequences of rule and theorem applications! Our excuse: we are trying to keep things simple.)

In particular circumstances we hope to arrive at a desired, expected or merely tentatively hoped-for conclusion and to do so via a sound, provable building process. We can liken it to using material to make bricks and other basic building blocks, using these to make sub-assemblies and components and so on. But to make what, ultimately? It depends upon context. It may be that component manufacture is where we stop, because we then sell components as products. Or it may be that we go on building until we have a very complex manufactured product such as a power station. Proof that the complex product will work as specified depends upon the quality of proof of each component and sub-assembly. The nature of this kind of proof is qualitative of course, so the analogy with logic is purely for illustrative purposes.

It would be nice to talk about computer-based information systems with more confidence than this. However, methodical structuring of the design of software is still theory rather than practice. There is still some way to go before we can have proof that a program is correct given the premise of the design specification and the languages in which it is written, though there is increasing optimism that it will be possible with certain types of software before too long. Safe defence systems may depend upon it!

For any particular argument we shall call what we are given to start with the *premises*. These may be axioms or theorems or both. Very basic arguments will start from axioms only. A single step of the argument is the transition from one or more of the premises to a new formulation, or from one or more previously derived formulations (plus none, one or more premises) to another formulation using an axiomatic rule of inference or one based on a theorem. Each step has to be justified by citing the rule used. Thus it goes:

Premises
Step: New formulation-1 (justification = rule + premis(es))
Step: New formulation-2 (justification = rule + premis(es)
                                    + formulation-1)
Step: New formulation-3 (justification = rule + premis(es)
                                    + formulations-1 and 2)

...

Step: New formulation-n (justification = rule + premis(es)
                                    + formulations-1 to $n - 1$)

until a step result is recognised as being what it was required to prove. We can then call this the *conclusion*.

Two problems come to mind immediately. Who is to audit the justifications? And how do we know that any step-by-step derivation is heading in the right direction towards the conclusion we require? The first problem is very difficult to answer in general. Though it may be possible for arguer and reader to agree each step plus its justification on its own merits, the setting of general standards for justification level and language is a thorny issue which is usually tackled pragmatically. Thus an expert system explaining how it has reasoned a response will have to have had a certain standard of quality of justification built in to it.

The 'which step next?' question is hardly a less-tractable problem. At any stage the number of feasible next steps is theoretically infinite (see next section), and in any case will be large even with the most simple of systems. There is no question of trying blind-search of all these, even using high-speed powerful computers. Heuristic search has to be the answer, and it is the method used almost invariably by human beings. This implies some next steps being more fruitful than others. Some may be obvious without looking at the required conclusion—rules of simplification in algebra are examples of this—others may only be candidates because of previous knowledge of similar required conclusions. (See the discussion in Section 6.4.5 on this.)

Finally, the step itself may be viewed as an application of a number of basic clicks of the logic turnstile:

from: formulation-n

to: $\vdash$ formulation-n

to: formulation-$n \vdash$ formulation-$(n + 1)$

to: formulation-$n \vDash$ formulation-$(n + 1)$(with appropriate justification)

to: formulation-$n \rightarrow$ formulation-$(n + 1)$

to: formulation-$(n + 1)$

7.2.3 The components of argument

We should now see that logical argument comprises a connected sequence of steps. At the start there exists a set of premises. It is an unordered set since the premises need have no particular sequence. Each step is the completion of a logical implication, the first having an antecedent which is any logical function of the premises, and all subsequent steps having an antecedent which is any logical function of: the premises *plus* the set of the consequents from all previous steps. This is shown schematically in Fig 7.2.

A number of observations can be made about this:

a) The actual step taken may only be one among a choice of many.

b) Each step requires some kind of justification.

c) Unless the argument is exploratory or random there must also be some goal so that there can be a concluding step which completes a required proof or answers a specific query.

Point *a* brings into play the important question of uncertainty about how to proceed with an argument. For example, suppose we have a set of premises which might be represented symbolically as (using comma separators in the list):

A, ˜C, A → B, D → C

There are a number of possible first steps. Firstly each separate premise and any conjunctive combination of them can be implied as being true in isolation. Thus we can clearly deduce that A is true, or that B is true, or that both A and B are true, ie. (A & B) is true. But we can also assert that ˜˜A is true and that ˜˜˜C is true, adding as many tilde pairs as we like and therefore generating an infinite number of possible steps on this criterion alone.

It is also possible to assert other, less obvious, equivalences. An implication is reversible if both sides are negated, for example. The premises A → B and D → C can therefore be reformulated as ˜B → ˜A and ˜C → ˜D (see Section 5.2.3), so each of these is also a possible first step.

And after these first steps we can add the consequents to our set of premises as material for the next step, thus rapidly expanding even further the possibilities for the next step. (Adding equivalent forms, such as reasserting that A is true for example, does not in fact extend the set of

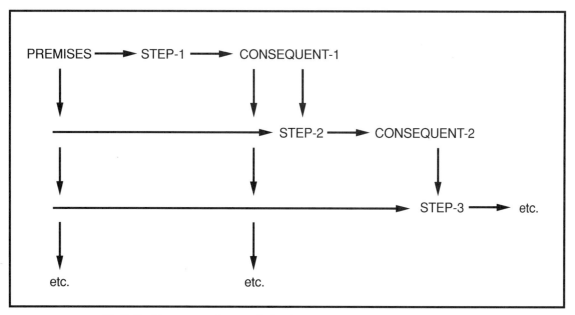

Fig 7.2 The steps of pure deductive argument

premises of course because it duplicates an existing premise, but it might nevertheless be necessary to make the reassertion deep into an extended argument.) So it should be clear by now that the 'next step' in any argument is not something which offers itself as inevitable and obvious.

Point *b* reminds us that when we say things like 'clearly' and 'we can assert', we are appealing to a common recognition of the rules of the game. A 'proper' argument will need to give a justification for each step by citing a previously established valid rule. For example, the justification of the reassertion of any premise or combination of premises lies in the rule

$$A \& X \to A$$

In formal logic such rules are given names and the name may thus be quoted in justification. However, all justification is relative. In the above, have we justified that A and X can represent one or more of the given premises? Does a list of premises actually justify a conjunction of all of the items on the list? It ought to, by the very nature of a premise being a premise. But any hidden contradictions among the premises could be disastrous. We can see that justification itself involves debate.

While logicians need to work this problem out with meticulous precision and clarity, we can take a more pragmatic view and say something like: "Whether or not I accept the justifying argument will depend on how I react at the time". When running an expert system the user should ideally have the facility for being able to query each step by asking Why? and repeating that question back into the argument until satisfied. Such explanations, although they may be shown to be logically sound and satisfying the audit of the professional logician, will always have qualitative and contextual significance. A 'good' explanation for one user may be incomprehensible to another. Indeed, there is a danger that rigorous-sounding explanations may conceal errors and flaws from the common-sense awareness of the user.

Point *c* suggests that it seems sensible to set argument in motion only if there is some agreed goal. Well, we could make the goal that all possible theorems be generated from a given set of premises, in which case the processing will be a kind of blind search of the logic. This is unrealistic in all but a few cases of pure mathematical research. Ideally, we would like in a two-valued system to be able to set the goal in two-valued form so that we could then add the goal to the set of premises and see what this does to the argument process when we trigger it into action. Such tactical considerations will be discussed in a later section (see Section 7.4.3).

7.2.4 Presentation of argument

A browse through any selection of texts with 'logic' in the title (or we could even broaden that to include 'mathematics') will present the browser with a potentially wide range of alternatives for the setting out of reasoned argument. Indeed this text is itself no exception. Narrative, evaluation of truth tables and line-by-line argument steps have all been used. In a sense it doesn't matter what the presentation looks like so long as the reader can

a) Be sure which formal system is being used.
b) Recognise the premises for any given argument.
c) Recognise what is the correct sequence of steps.
d) Accept the justification given for each step.

Point *a* is very context-sensitive. An argument presented within the pages of a specialised textbook will need no statement about the formal system if that is what the text is about anyway. Where a text is covering a number of different systems (as is the case in this text) greater care is needed and some confusion is almost inevitable at some point, especially if a reader 'dips into' or 'flits about' the pages of the text.

In the context of running software in a mixed environment, point *a* is tested by what the user sets up in terms of operating system and packages before running a program. As systems become more genuinely distributed and as packages are made to be compatible with other similar packages, this will surely become a problem akin to the dip-and-flit reader mentioned above but with the much more serious danger of some erroneous action being taken in consequence. Human computer interfaces (HCI) need to be designed with sufficient intelligence to guide and teach the user and, if necessary, to prohibit the user when imminent wrong action is perceived.

Point *b* concerns certainty about the set of premises. Mathematicians will usually make an explicit prefix such as 'Given that...' and logicians might use $\vdash$, though this needs to be used with some careful note of context as we have already discussed.

On the printed (or written) page it should not be difficult to spell out what this set is by some kind of listing or tabulation. Premises are necessarily all true *at the same time* and therefore the comma or space separators of a horizontal list or the new-lines of a vertical list or tabulation in this context represent logical ANDs. This might be literally the meaning in some kinds of software interface. In any case, relatively intelligent software should allow the user to specify premises as naturally as possible while not permitting different-time premises to be merged unless they are deliberately meant to be.

Point *c* concerns the essential requirement of making clear the step-by-step turnstile clicks of the argument. The almost universally adopted scheme is to present the steps as a top-to-bottom line-by-line list. If premises are also listed in this way, then there needs to be some explicit labelling to separate the lists visually.

In some circumstances it may be useful to use a kind of flow diagram, although flow diagrams are more usually associated with the presentation of programming logic and data flow logic. Both these logics are to do with action sequences rather than the formal logics we are concerned with here. The steps are literal statements about what happens next as a matter of fact rather than of logical necessity—instruction statement Y has been written to follow instruction statement X in the control sequence because the programmer wants it to. Though design choice will have been constrained to some degree by the system designer's specification, this kind of logic is very different in nature to the logic of argument.

The software of an inference engine will of course get on with the processes of argument hidden from human view and it will 'know' the correct sequence by virtue of its own internal system design. However it may be asked by the user to present part or all of its reasoning and therefore the interface designer has to tackle the issue of argument presentation as a screen-display and/or printed listing.

We have dealt with point *d* as best we can in earlier discussion. The issue of what constitutes satisfactory justification is not simply a question of rules of logic but is strongly affected by the nature of the user-system relationship (or reader-writer relationship in written presentations). This is such a context-sensitive situation that we can do no more here than urge that the system designer be aware of its importance, and do all possible to make human-computer interfaces user-friendly yet intelligently safe from error. This is no mean challenge!

7.3 Rules of inference

7.3.1 Introduction

As users of expert systems or even as designers of expert systems using shells or other design aids, we do not really need to be adept at operating the rules of logic. The software will in general take care of any need to do complex sequences of inferencing. However, it does give an analyst a certain extra confidence if there is some understanding of what goes on in the inference engine-room. There are also some aspects of the limitations of what can be achieved which need to be understood and kept in mind by the designer.

We have already introduced the basic process of inference in the previous sections. Let us reiterate first, though, the important fact that we are dealing with material already held to be true when we operate the logic turnstile. The truth table by its very nature seeks to look at all possible combinations of the truth values of component variables and this can sometimes lead to awkward not-specifiable results as we have seen. If we stick only with rules that take us forward through the argument, we can avoid these awkwardnesses to some extent.

The rules of inference that we shall look at in this section present the reasoning machine with various options when it comes to trying to take an argument forward. Whether a particular machine uses only some of these or not will depend upon the design of that particular software. The need for heuristics quickly becomes apparent as we look at the characteristics of each rule.

7.3.2 Modus ponens

The use of Latin for this propositional method is an indication of the classic vintage of the rule. It is indeed the rule we used in a previous section to illustrate the main step of an argument. It is of the form

Given: implication A → B is TRUE
Given: proposition A is TRUE
Deduce: proposition B is TRUE

It is hoped that we do not need to say any more by way of proof that this rule does work. We have explored it fairly thoroughly and have seen that it satisfies the strict Boolean requirements of the implication connective. But note how each line is an assertion that something is TRUE. Once it is understood that we are working in this mode we do not need to keep adding that phrase to the end of each line, but it is just as well not to forget its implicit presence always because it is telling us that *we are dealing with less than the complete set of rows of the truth tables*.

This is further emphasised if we look at the logic map illustration of the rule as shown in Fig 7.3 (This first appeared as Fig 3.9(ii) in Section 3.3.2, together with an explanation.) Because A → B is TRUE, the diagram has to be special, with the area A inscribed within the area B. Certain types of area simply do not exist (A & ~B in this case).

From an inference machine point of view we can deduce that it can make use of some heuristic to the effect of: Search for a true assertion which is itself an antecedent in a true implication (ie. a given IF... THEN... rule, for

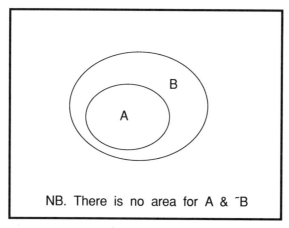

NB. There is no area for A & ~B

Fig 7.3 The sets illustration of A → B again

example) and then apply mod pons. (The abbreviation 'mod pons' is fairly commonly used.)

The symbol A can itself represent a more complex expression. This is the case in a decision table, where A represents the conjunction of all the correct condition values and B represents the corresponding action(s) in the rule column. If an expert system is designed entirely on the basis of a decision table structure, then the heuristic is the matching of given conditions to the condition quadrants of the tables, though there will also need to be some preliminary heuristics on how to pick out the right table of course.

7.3.3 The chain rule

The importance of the chain rule is that it takes an argument forward by generating a new true *implication*, rather than forward to just a new proposition as is the case with mod pons. It is also very easy to accept as a valid rule by using plain common sense. It has the form:

Given: A → B
Given: B → C
Deduce: A → C

EXAMPLE   If a Form is marked L then it is used in the laboratories. All Forms used in the laboratories are liquid-proofed. We can deduce therefore that all Forms marked L are liquid-proofed.

(Note that these are indeed implications, not equivalences, since there may be other Forms than those marked L used in the laboratories and liquid-proofed Forms used in places other than the laboratories.)

It can be shown by full evaluation of a truth table for the expression $((A \rightarrow B)(B \rightarrow C)) \rightarrow (A \rightarrow C)$ that it is tautologous. However we can prove it more concisely by the use of assumption, as follows:

Given: A → B
Given: B → C
   Assume: A
   Deduce: B   (by mod pons)
   Deduce: C   (by mod pons)
Deduce: A → C   (since A true led to C true)

The assumption section of the argument has been indented to clarify that it is a sub-argument. A is not *given* as true and must not be part of the main argument. Mathematicians, especially geometricians, use *assumption* extensively in their proofs. The last line is actually justified by the rather more formal theorem that $X \vdash Y$ is equivalent to $\vdash X \rightarrow Y$.

The special Venn diagram illustration of the chain rule is even more explicit (see Fig 7.4). (This was used as an illustration in Section 5.2.3.) Since A must be inscribed within B (ie. A → B) and B in turn within C (ie. B → C) then A must be inscribed within C (ie. A → C).

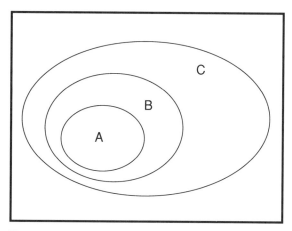

Fig 7.4 The chain rule illustration again

The heuristics which would seek to make use of the chain rule will obviously be based on finding implication pairs with the consequent of one matching the antecedent of the other. The successive application of the mod pons heuristic will of course achieve the same effect as the chain rule, but if the chain rule heuristic is applied independently then it can potentially generate many new implication rules to extend the rule base in readiness for the feeding of conditions to the system. There will however need to be another heuristic checking whether such unlimited rule generation is viable since it might in practice cause the workspace to overflow!

7.3.4   The resolution rules

Resolution presents us with a third type of inference rule, namely one which generates a new expression from two given expressions. It has the form:

Given: X ∨ A
Given: Y ∨ ˜A
Deduce: X ∨ Y

EXAMPLE   Either File X is open or User A is on-line (or both) and either File Y is open or User A is not online (or both). We can deduce therefore that either File X or File Y is open (or both are).

(Notice how difficult this is to follow and therefore also presumably to accept. The fact is we probably wouldn't have set out the premises in this way at all! The reality is that one or both of the two files is always on-line and this is because File X must be open if A is not on-line and File Y if A is on-line. Sort that out and you will see that resolution is really the chain rule in disguise! See below.)

The rule was shown in this form in order to point up the useful-looking *elimination* feature, whereby a common variable in two expressions can be dropped. It is useful in the practical sense of providing a simple heuristic, especially when its even simpler form is included:

Given: A   (A is on-line)
Given: Y ∨ ˜A
Deduce: Y   (File Y is open)

because if all the given formulae are expressed as disjunctive clauses then the heuristic is a search for any common variable terms with complementary form (A and ˜A) so that they can be chopped. Remember that X and Y can themselves be disjunctive clause expressions (disjunctive normal form), so we are really asking that all the given formulae be expressed in DNF which again is a standardisation heuristic likely to be very effective in designing reasoning machines. There is even greater significance to all this as we shall see later.

The second form above reveals fairly clearly that we are actually dealing with a thinly disguised form of mod pons, if we recall that the Boolean algebra equivalent of $A \rightarrow X$ is $\tilde{A} \vee X$! What then is the full form of resolution that we started with? It is none other than:

Given: X ∨ A   which is ˜X → A
Given: Y ∨ ˜A   which is A → Y
Deduce: X ∨ Y   which is ˜X → Y (the chain rule!)

So the resolution rule has the great merit that it incorporates both of the powerful inferencing rules of mod pons and the chain rule and operates in a very tidy way.

7.3.5 Other rules of inference

There are of course a great many other elaborations of the basic rules which throw up other potentially useful rules. However, the resolution rule tends to show itself to be a more practical equivalent for many of these. But when reasoning 'on paper' it is often easier to work with the more graphic chain rule and therefore the following rule (cf. Section 5.2.3) is actually very useful:

Given: A → B
Deduce: ˜B → ˜A

In DNF form it is simply

Given: ˜A ∨ B
Deduce: B ∨ ˜A

which will show that it is in fact an equivalence, merely a restating of the formula. The turnstile will go in both directions with this step which is just a practical application of the commutative law. Nevertheless, it can be useful to doodle with the reversing of implications in this way when working by hand.

EXAMPLE Required to deduce X given the following as premises:

$$\tilde{A} \to X, \; D \to P, \; Q \to D, \; A \to Y, \; P \to \tilde{Y}, \; Q$$

By spotting the matching consequents and antecedents we can juggle these into a chain rule argument as follows:

$Q \to D$	(given)
$D \to P$	(given)
$Q \to P$	(chain rule)
$P \to \tilde{Y}$	(given)
$Q \to \tilde{Y}$	(chain rule)
$A \to Y$	(given)
$\tilde{Y} \to \tilde{A}$	(equivalence)
$Q \to \tilde{A}$	(chain rule)
$\tilde{A} \to X$	(given)
$Q \to X$	(chain rule)
Q	(given)
X	(mod pons)

For comparison let us set this out as a resolution-applying exercise. Expressing all premises in DNF, we have

$$A \lor X, \; \tilde{D} \lor P, \; \tilde{Q} \lor D, \; \tilde{A} \lor Y, \; \tilde{P} \lor \tilde{Y}, \; Q$$

While a machine could be left to rattle through all possible resolution eliminations until a formula X pops up, it is rather more difficult by hand since the chaining direction of the arrows is lost. You should try this for yourself.

7.4 Proofs and methods of proving

7.4.1 What is a proof?

This is perhaps an appropriate point at which to say something once more about formal systems. The ideal formal system is one which clicks away on its own turnstile in an entirely self-sufficient way—the science fiction fantasy of a perfect and even god-like mind-machine. But it is like the dream of the perpetual motion machine or the endless steps of an Escher tower—the almost graspable idea turns out to be impossible in the real physical world, for energy conservation reasons in the first case and space-time reasons in the second. The perfect mind-machine system founders in the shadowy area "between the idea and the reality" as the poet Eliot had it, because every idea has to pass from one mind to another through the real world.

Every system, however well formed, consistent and apparently complete, requires at some stage of its implementation or use to be described. The description itself has to be some kind of language system which in its turn will need a description language and so on ... ad infinitum? In practice, philosophers, mathematicians, writers all call a halt to this endless wrapping exercise quite soon and accept that 'informality' *must* be allowed in if one thinker is to communicate with another.

There is more to it than this though—much more. There are some parts of any formal system that cannot be handled by the system itself! This is the crux of Goedel's Incompleteness Theorem published in 1931 when discussing the work of Russell and Whitehead. Hofstadter first published his entertaining expansion on this and other issues in 1979, using the epithet Strange Loop to cover these undecidable phenomena in the different fields of *Goedel, Escher and Bach* as his title elegantly expressed it.

Since we are here concerned with logic, let us look at the most famous 'liar' example. Consider the statement:

"I am a liar."

This statement, let us call it A, implies that all statements made by me are FALSE. Put in blunt symbolic terms this is

$A \rightarrow \tilde{}A$

Purely formally then, this can be written

$\tilde{}A \vee \tilde{}A$

which is $\tilde{}A$. Therefore A must be FALSE. I am not a liar. But then again I must be because, if I'm not, my statement A is TRUE, and I *am* a liar!

The issue will remain *undecidable* because it is an example of the part of a formal system that the system itself cannot handle. To get anywhere with it we should for example have to start suggesting that if I understood logic I wouldn't make such a silly statement. Formally, we could try banning $A \rightarrow \tilde{}A$ from the system, but that just leads to further trouble of a similar kind. *Self-reference* as it is called is the cause here, but we shall have to beware it rather than ban it. For a start we might say that the sentence "I am a liar" is not the same as the fact that I am a liar—they belong to different systems. At least it will prevent that particular nasty from popping up in future. But the very issuing of this warning has had to be put in an 'outer-casing' language please notice.

All this is crucial to the question of *what makes a proof*, because a proof—a sequence of justified steps that takes us from a set of premises to some concluding step—has to resort to an outer or higher system to present itself. A suitable homily for the systems analyst/designer here would be to point out that, no matter how perfectly tested the software system, if the user makes an unexpected error then error there is in that more complete system of software-plus-user. And who can claim that the software untouched by user hand is a system at all in any useful sense? Conclusion—the system can never be fully tested.

We have been presenting proofs fairly informally anyway in this text. More formal texts will be found to have some difficult abstract sections to try to deal satisfactorily with the issue. We hope here to have at least raised the level of awareness of the practising systems analyst that such philosophical issues are not as remote from their work as they might have believed.

7.4.2 Methods of proving

All proof works through the general theorem about proofs that we have already made use of; this says that

$$A \vdash B \quad \text{is equivalent to} \quad \vdash A \rightarrow B$$

This tells us that if you can derive B from a starting premise A, then that is tantamount to being able to say that you have proved that the truth of A implies the truth of B. This is as abstract a view of the matter as we shall want to take, and we shall leave the theorem, an eminently common-sensical one, to speak for itself.

Instead we move on to the practical methods of achieving a satisfactory proof. First, it has be said that particular methods tend to be associated with particular formal systems. There is, for example, a wonderfully elegant *proof by induction* which is particular to sequential systems, especially the system of natural numbers. We outline it out of interest.

- Required to prove some law of number true for all natural numbers.
- Show that it is true for some low-valued integer 0, 1 or 2.
- Assume it to be true for a general case n.
- Substitute $n + 1$ for n in the law and show that the law still holds true.
- It is therefore proved that the law holds for all natural numbers from the low-valued starter upward.

This works because the step from n to $n + 1$ is perfectly general to the whole sequence of natural numbers.

EXAMPLE   A rule which is connected with the graph theory we discussed in Chapter 4 (4.6.2) is also quite useful in relational analysis. It states that the number of possible pairings between any number of entities n is given by

$$n * (n - 1)/2$$

(We are using the asterisk for multiplication and slash for division.)

Pairing makes no sense for one entity but we know that the answer is 1 pairing for two entities and can test that the law gives this answer for $n = 2$:

$$\text{Number of pairings for } n = 2 \text{ is } 2 * (2 - 1)/2$$
$$= 2 * 1/2$$
$$= 1$$

Now assume the law is true for the general value n and then introduce a new entity. This introduction must produce n additional pairings, one with each of the existing entities. So we add n to the number of pairings for the general case as follows:

$$(n*(n-1)/2) + n$$

and now try to show that it gives the right formula in terms of $n + 1$:

$$
\begin{aligned}
&= (n*(n-1)/2) + 2*n/2 \\
&= (n*(n-1) + 2*n)/2 \\
&= (n*n - n + 2*n)/2 \\
&= (n*n + n)/2 \\
&= n*(n+1)/2 \\
&= (n+1)((n+1)-1))/2
\end{aligned}
$$

To see that this is indeed the correct formula for $n + 1$, substitute N for $n + 1$, which gives

$$N(N-1)/2$$

By induction, we now assert the law to be true for all values of n from $n = 2$ upward. (It is always satisfying to try out a law for some actual values and Fig 7.5 shows it is true for $n = 5$.)

Understandably, number theorists set great store by this method of proof

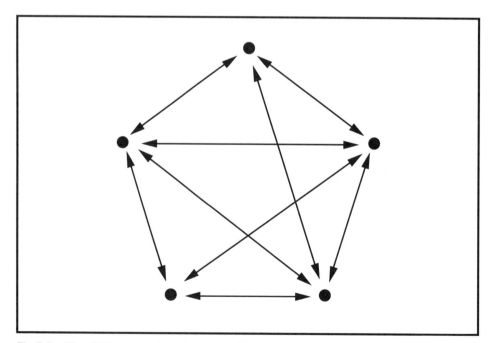

Fig 7.5 The TEN connections between FIVE entities

by induction, but unfortunately it cannot be transferred to simple logic. (Though it does have application, not surprisingly, in formal systems of list processing, since lists are sequentially ordered sets.)

The only method we have adopted so far in this chapter—and rather informally at that—has been to set the logic turnstile in operation and hope that we get where we want to. When working by hand we may rely upon keeping an eye on the required result and probably mentally working forward from the premises and backward from the required conclusion simultaneously. It can work well, particularly with those who are specially gifted at mathematical puzzle-solving, but it is not a good heuristic because it would be very difficult to clearly specify it in such a way that others (and especially a machine) could follow it.

A method that suffers rather less from this deficiency is the method of proof by contradiction, or, to give it its Latin name, 'reductio ad absurdum'—literally, 'derivation of an absurdity'. This is described in the next section.

7.4.3 Proof by contradiction—reductio ad absurdum

Suppose we are trying to prove that some result R can be derived from some set of premises P. If we adopt an assumption that R is in fact not true, ie. assume ~R and bring it into the process, it should generate a contradiction if R really was implied by the premises. In its simplest form a contradiction is given by

$$X \& \sim X$$

For the justification of this proof method we could appeal to the equivalence:

$$P \rightarrow R = \sim P \vee R$$
$$= \sim(P \& \sim R)$$

The final expression is arranged in a form which tells us that P and ~R being true together would give a FALSE result if $P \rightarrow R$ is indeed TRUE. The justification is to be read as being expressed in an outer-case language, but it happens that we can run this language in Boolean form.

The great advantage of the method in inference machine design is simply that it can be very effective. Rather than setting the machine off in search of a specific formula result, we can prime it with the negation of the result and then set if off to trip over the contradiction which we suspect is now inherent in the premises with which it is working.

EXAMPLE Prove that $R \vee S$ can be derived from the set of premises $P \vee Q$, $P \rightarrow R$, $Q \rightarrow S$.

[Sample paraphrase of the problem: We must have in this new package either high productivity (P) or quality assurance (Q) or both. High productivity gives good Rate of Return (R) and Quality Assurance guarantees Customer Satisfaction (S). Therefore we are saying that we want to guarantee good Rate of Return or Customer Satisfaction or both. Is this sound logic?]

Assume ˜(R ∨ S) and add it to the premises, then search for a contradiction.

We shall use resolution and therefore need to express all premises in DNF, as follows:

P ∨ Q		already in DNF
P → R	gives	˜P ∨ R
Q → S	gives	˜Q ∨ S
˜(R ∨ S)		
= ˜R & ˜S	gives	˜R, ˜S (two premises)

The advantage of having ˜R and ˜S as premises is their ability to eliminate from other expressions by resolution.

˜P ∨ R, ˜R	resolve to	˜P
P ∨ Q, ˜P	resolve to	Q
˜Q ∨ S, ˜S	resolve to	˜Q

The last two steps result in the contradiction we are seeking. We can say we have proved that R ∨ S is true by contradiction.

Two cautionary notes must be sounded about this method. The first is that a system which is in a contradictory state as it were must be used for nothing else than seeking the contradiction. The reason is that contradiction has the unfortunate characteristic that it implies that anything and everything is true! We can see this from the Boolean interpretation:

$$(A \& \tilde{\,}A) \to X$$

(where X is anything we want it to be)

$$= \tilde{\,}(A \& \tilde{\,}A) \vee X$$
$$= (\tilde{\,}A \vee A) \vee X$$
$$= T \vee X$$
$$= T$$

Tautology. Therefore the first line is always true.

This gives imaginative logicians a field day thinking up examples which are logically true but patent nonsense from a common-sense viewpoint. A business-related example might be: The company made a profit last year and yet did not make a profit, therefore all employees can take a month's holiday. The god of logic might say: "You are trying to work with and make some sense of a total contradiction! You deserve all the nonsense you get!" At a more practical level it does mean ensuring that our strictly logical inference engine is designed so as to prohibit being interrupted for interrogation while a proof by contradiction is active, unless of course it has the appropriate routines for freezing and setting aside such working while it attends to the interrogation.

The second cautionary note concerns the situation where proof by contradiction is being attempted when that which is to be proved is actually not true. The search for contradiction is almost limitless—not quite limitless, because the engine could recognise the point when it has tried all possible derivations and still not hit a contradiction and then stop. But the number of possibilities to be so tested could be so great as to, in effect, take an infinite time to explore. Obviously there need to be time limits set on such searches. A software engineer would probably recommend that time limits be set in any case for any open-ended or blind searches.

7.4.4 Proof by evaluation

We have used the property that a tautologous outcome is proof that we have started with a valid set of propositions. If we can express an argument algebraically and evaluate it to such an outcome, either by algebraic manipulation or by truth table evaluation, we shall have proved the argument to be valid. Both of these approaches may be tedious in practice, and again the reductio ad absurdum approach points to a more effective short-cut. If we can show that for just one set of values of the variables that the premises are true while the result being tested is false, then we have shown that the result cannot be sustained at all. So again if we set up the negation of the result to be proved and achieve the above then we have our required proof by reductio ad absurdum.

The 'for just one set of values' implies testing just one row of the truth table, but on the other hand assumes that we can alight upon that set immediately. Let us for example set out to prove the argument that if $A \rightarrow (B \& C)$ and $B \rightarrow A$, then $B \rightarrow C$. An actual example might be:

> If we hire extra staff (A), then we shall meet the deadline (B) and it will cost more (C). [This is $A \rightarrow (B \& C)$.] We know that meeting the deadline means hiring extra staff [$B \rightarrow A$]. So meeting the deadline will cost more [$B \rightarrow C$].

We need to find a set of values of A, B and C which satisfy the premises but make $\tilde{}(B \rightarrow C)$ false. We *might* spot immediately the case of all of A, B and C being true as meeting this requirement! It is fairly obvious that these values satisfy the two premises $A \rightarrow (B \& C)$ and $B \rightarrow A$ and also the implication $B \rightarrow C$. But if $B \rightarrow C$ is true then its negation is false, which proves that the argument must be true for all sets of values.

This 'trick' won't always work so sweetly. Remember that there are fifteen other value sets and in some cases it might take tries at all of them before we hit on the crucial one, which is nothing else but full truth table evaluation after all. The method does provide an option to try however.

7.5 Clausal forms

7.5.1 A reminder about normal forms

We introduced in Chapter 3 the two normal forms of Boolean algebra—the conjunctive and the disjunctive. Any expression can be re-formed so as to be in either of these two forms.

The conjunctive normal form of clause-strings, connected by the & connective

$$C_1 \& C_2 \& C_3 \& \dots \& C_n$$

where the clauses C have no embedded & in them, has the properties that

- All clauses must be 1 (TRUE) to make the whole expression 1 (TRUE).
- The whole expression will be 0 (FALSE) if any one or more of the clauses is 0 (FALSE).

There are dual properties for the disjunctive normal form. This is the form of clause-strings connected by the ∨ connective:

$$C_1 \vee C_2 \vee C_3 \vee \dots \vee C_n$$

where the clauses C have no embedded ∨ in them. It has the properties that

- All clauses must be FALSE (F or 0) to make the whole expression FALSE.
- The whole expression will be TRUE (T or 1) if any one or more of the clauses is TRUE.

These properties are useful in considering a certain model for arguments as we shall now see.

7.5.2 Normal forms in argument

A *clausal form* of an argument is defined as being of the general shape:

$$(C_1 \& C_2 \& C_3 \& \dots \& C_n) \rightarrow (D_1 \vee D_2 \vee D_3 \vee \dots \vee D_n),$$

where the consequent clauses have been labelled D to distinguish them from the antecedent clauses C. We have seen that for any deductive step in an argument a certain set of premises are true at the same time, from which can be derived the next formula, and that in general a number of possible derivations are available from that set. The clausal form epitomises this in terms of normal forms.

In a sense it summarises in one formula the whole of the possibilities (the D clauses) of any set of premises (the C clauses). But its generality makes it very difficult to imagine implementing as a universal formula for an inference engine, since, as we have already discussed, the real problem is finding the right D or combination of them at the right time. In large-scale arguments the search could take forever.

We get a little nearer the step view of argument if we consider the simpler model as having only one consequent clause:

$$(C_1 \& C_2 \& C_3 \& \ldots \& C_n) \to D$$

which is after all like a single step of any argument. When additionally we constrain each of the clauses C to being atomic, which is a precise way of saying that it comprises only a single proposition or negation of a proposition, this is called a Horn clausal form, or simply **Horn clause**, named after Alfred Horn's work in 1951. It has the remarkable property that any form of implicative expression can be re-formed to be a Horn clause or conjunction of Horn clauses.

Since the Horn clause is very simple in structure this is a very important discovery and has been put to direct practical use in logic-driven languages such as Prolog, which we shall look at in the next chapter. We shall conclude this section by showing briefly that the claim we have just made is true.

7.5.3   Reducing all things to Horn clauses

Let us make things simpler for ourselves by seeking to show that all is reducible to simple Horn clausal form:

$$A \& B \to C$$

We must first decide whether formulae not involving implication can be accepted as Horn clauses. The single atom A can be dealt with by observing that A is the same as $A \vee F$, which is $A \vee {\sim}T$. So this is expressible as the implication:

$$T \to A$$

This is like re-phrasing the statement that 'A is TRUE' by something like 'that A is TRUE is always to be implied as being TRUE'. The complement ${\sim}A$ can be treated in a similar fashion:

$${\sim}A = {\sim}A \vee F = A \to F$$

This is like re-phrasing 'A is FALSE' by 'that A is TRUE always implies a contradiction'.

Let us first examine whether the form:

$$X \vee Y \to C$$

can be re-formed as Horn clauses. We shall use Boolean algebra to achieve this. We can rewrite the implication as

$$\tilde{\ }(X \lor Y) \lor C$$
$$(\tilde{\ }X \,\&\, \tilde{\ }Y) \lor C \qquad \text{(de Morgan)}$$
$$(\tilde{\ }X \lor C) \,\&\, (\tilde{\ }Y \lor C) \qquad \text{(factoring)}$$
$$(X \to C) \,\&\, (Y \to C)$$

And as long as we accept that conjunctions can be presented as a list, we have

$$X \to C \quad \text{and} \quad Y \to C$$

which are two Horn clauses.

Figure 7.6 shows this quite explicitly if we chose to accept this illustrative way of deriving the result.

Consider an actual example of this. Suppose the statement is made that

Improvement of sales (X) or reduction in costs (Y) will contribute to financial success (C).

Our conversion to Horn clauses tells us that we can deduce from this that both of the following statements are true:

- Improvement of sales (X) will contribute to financial success (C).
- Reduction in costs (Y) will contribute to financial success (C).

As with any actual example, though, care has to be taken to ensure that the specifically logical interpretations only are intended by the speaker. If the intention in the original statement is to say "*either* sales improvement *or* cost reduction will contribute, but I'm not sure which" then of course the inclusive OR cannot be used to represent this.

Now, suppose that we start with

$$A \,\&\, B \to C$$

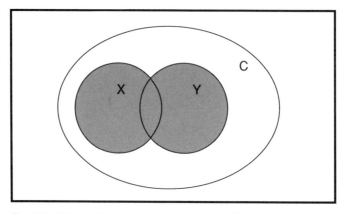

Fig 7.6 Illustrating how X → C and Y → C must be TRUE if (X ∨ Y) → C is TRUE

and discover that A is not atomic but is in fact, say, $A_1 \vee A_2$. We now have

$$(A_1 \vee A_2) \& B \to C$$
$$(A_1 \& B) \vee (A_2 \& B) \to C \qquad \text{(factoring)}$$

So we have the two Horn clauses:

$$A_1 \& B \to C$$
$$A_2 \& B \to C \qquad \text{(from our first result above)}$$

An example of this might be (slightly adapting our earlier one):

> Given that the market is buoyant (B), improvement of sales (A_1) or reduction in costs (A_2) will contribute to financial success (C).

The Horn clauses for this would be

- Given that the market is buoyant (B), improvement of sales (A_1) will contribute to financial success (C).
- Given that the market is buoyant (B), reduction in costs (A_2) will contribute to financial success (C).

When the right-hand side is itself a compound expression the re-formation is more difficult, but still achievable. We shall consider the general form:

$$E \to (A \& B) \vee (C \& D)$$

By factoring we can get

$$E \to (A \vee C) \& (A \vee D) \& (B \vee C) \& (B \vee D)$$

By reversing the implication and negating both sides we arrive at

$$(\tilde{}A \& \tilde{}C) \vee (\tilde{}A \& \tilde{}D) \vee (\tilde{}B \& \tilde{}C) \vee (\tilde{}B \& \tilde{}D) \to \tilde{}E$$

This again can be presented as a list of Horn clauses using our first result:

$$\tilde{}A \& \tilde{}C \to \tilde{}E$$
$$\tilde{}A \& \tilde{}D \to \tilde{}E$$
$$\tilde{}B \& \tilde{}C \to \tilde{}E$$
$$\tilde{}B \& \tilde{}D \to \tilde{}E$$

While Horn clause formation is a normalising process rather than a simplification of formulae, it does have an important role in preparing argument material for application of the Resolution rule, since every formula is guaranteed to contain at least one atomic clause, namely the consequent, which is then at the ready for cancelling its complement in other formula wherever it can be found.

7.6 Concluding remarks

In this chapter we have been looking at the simplest and soundest model possible for automating a type of reasoning process. It is some kind of foundation, but is only a modest achievement in many respects. The premises, the basis of the knowledge in our knowledge base, seem to be constrained to being expressible in two-valued form and the connective rules between data have to be expressible in the IF...THEN... form.

If we insist upon using the digital computer as the principal machine for implementing knowledge processing then it is true that ultimately the machine itself will have to do this anyway—working as it does as a mechanistic manipulator of binary digits. But we know that in practice we hardly ever these days concern ourselves with bits, binary arithmetic or symbol tables unless we are designers of competitively efficient software for specific machines. We are more likely to want to work with menus, with screen icons, with software tools related to our needs as system designers be they CAD or CASE, or with Expert System shells. Are these facilities merely pragmatically built aids or are they also related to the principles of logic and the design of knowledge representation? We shall try to address such issues in the final chapter of the book after having looked first at some other problematical issues in the intervening chapters.

8 Predicate Logic

8.1 Formulating predicate logic

8.1.1 About predicates

The **predicate** is a term used in grammar. We start to build a sentence in a language by having some subject noun or pronoun and adding a predicate to it—a qualifying expression that tells us something more specific about that subject. In this, its grammar usage, the predicate might be as simple as a single verb:

The program [subject] works [predicate].

At the other extreme, being everything other than the subject, the predicate might be a complex grammatical structure involving clausal structures:

The program [subject], written by our most experienced Cobol programmer and completed only yesterday, works when tested on the new computer installed last week [predicate].

So the predicate is a rather broad term used to mean anything which enlarges upon, or provides more precise information about, the subject. It is the spirit of this interpretation rather than the literal meaning of it that is carried over into logic (and also into some programming languages, as we saw in Chapter 5).

A simple logic statement or proposition is, as we have seen, one which must be capable of having only two values, TRUE or FALSE. There can be no values in between—hence the name given to this constraining rule: *the law of the excluded middle*. Little scope there then for predicating, unless we consider moving on to other than straightforward two-valued systems.

However, there is another way in which we can predicate a proposition and that is by investigating whether the subject of the proposition might be identifiable as a set. The Venn diagram has already given us a clue to this possibility, since each circle can be interpreted as enclosing a set or class of objects or elements. The intersectional area for two overlapping circles represents the logical AND and can be interpreted as meaning those elements which belong to both sets or classes. And so on as we have seen for other logical connectives.

We could explicitly predicate a proposition about a set of objects by showing these in the label for the proposition. For example, take the very tangible set—a file of records. To express this in one proposition as to the fact of the file's existence we could write F(r), where F symbolises the file and r symbolises any record on the file. Actually, since F(r) is to be a two-valued proposition, r can be any record which might be a candidate for being on the file, so that should we have a specific record r_1 to discuss we could assert that F(r_1) is TRUE if that record is on the file and FALSE if it isn't.

We shall define this convention more carefully shortly, but before losing the thread of the idea of predicating F let us consider a slightly more complex set. Supposing each record has a type code t associated with it, perhaps defining its security level for access control. We could capture this by attaching two predicate classes to F by writing F(r, t). If r_1 is a particular record with a type code of t_1 and it is on the file then F(r_1, t_1) is TRUE. This begins to look very much like the notation for relations which we met in Chapter 5. This is no accident, for relations do represent a special case of predicating, and the logical calculus associated with them tends to be special too as we saw. The need to be able to interrogate relational databases using inference engines in the future will mean that some specially adaptive interfaces will have to be developed. However, for now we will continue to study the more general case of predicated logic to see where it gets us.

8.1.2 Domains

Although in files and databases the sets of data are finite, the logic assumes initially that the symbol(s) within the brackets representing the predicate(s) can represent classes of indeterminate size. This means that our understanding of the logic of sets as we discussed it in Chapter 5 can be brought over into predicate logic, though, as we shall see, one or two of the concepts have to be worked through again in the light of the new context.

It is easier for us to understand how the predicate logic works in the information processing arena if we confine ourselves deliberately to finite sets, that is to finite numbers of elements within classes. This is narrower than the Universal set concept, since we need this boundary for each class and we give them a new name—we call them **domains**. We met this term first when discussing relational operations, and made the comment that it was a troublesome concept.

Take, for example, a company personnel database comprising one record for each employee. The actual records represent the TRUE values of the database proposition, but a predicate symbol representing 'any' individual will cover all 'employable' people—it is simply that the proposition will be FALSE for any individual not currently on the database. Such a domain could be indeterminate if the membership criterion is 'if it's human, it's employable'! On the other hand if the domain is defined as being only those records actually on file, then there should never be any instance of the proposition being FALSE. This represents tight control and may be regarded as a good thing, but in the wider usage of knowledge processing the constraint on intelligent search of all data could prove over-restrictive—a response of 'never heard of him', as it were, to a query about an employee recently retired would not be regarded as being very helpful.

In practice the domain could usefully comprise past and present employees, with the possibility in larger organisations of including current applicants for posts and in some types of contractual industry of including people who are on the books as potentially contractable. Experienced data processing practitioners will recognise in all this the approach of keeping records on different files according to their status (or of giving them different status codes if all in the same database). Thus a record might go from an Applicant status to an On-strength status, and eventually to a Retired or Archive status. At worst, each of these different conditions might be remote from one another, both in the sense of being on different physical files (and even an efficient database is not going to keep all archived records at the same physically accessible level as the current records) and in that they change in format as they go from one status to another. It will be difficult to ensure that there is software that is aware of all these as belonging to a common domain if software suites have been developed separately for each view of an employee.

So perhaps it is not the concept of domain that is troublesome—it is not difficult to see what it should mean in principle—the problem is the actual defining of it in practice for each and every data class so that software is intelligently aware of the domain. Predicate logic starts at this wider perspective and therefore offers a potential solution.

8.1.3 A notation for predicate logic

By combining our logic notation with the relational we should arrive at the right type of notation for predicate logic. Rather than rigorously covering the same ground again, we shall present the notation in ready-to-use form and develop understanding through examples. In order to emphasise its use in getting a little closer to modelling natural language we shall also use more explicit names for both the propositions and their predicates.

For example, let us start by making a few assertions about employees of a fairly general nature and seeing how these might be modelled in the notation.

Consider first: "All employees who are on-strength pay National Insurance contributions".

$$\forall \text{ employee. ON\_STRENGTH(employee)} \rightarrow \text{PAY\_NI(employee)}$$

We could paraphrase the notational model as: "For every employee for whom it is true that they are on-strength, it must also be true that the employee pays NI". This is recognisably the same as the original proposition, if necessarily a little clumsier in style in order to match it to the model. (Turn back to Section 5.3.1 for revision of the meanings of ∀ and ∃.)

Consider next: "There is at least one on-strength employee on the Board." This might be modelled:

∃ employee. ON_STRENGTH(employee)
& BOARD_MEMBER(employee)

Again paraphrasing: "There is at least one employee for whom it is true to say that (s)he is both on-strength and a member of the Board."

Now let us assume that employee is the name of an entity whose elements are actual names and initials, such as Smith A. E. (It would be more likely that these were employee reference numbers, but the names make the example easier to read.) Because employee can take on any of these values it is of course a variable name, and could have been represented by any symbol so long as the same symbol appeared through the formula. For example,

∀x. ON_STRENGTH(x) → PAY_NI(x)
∃x. ON_STRENGTH(x) & BOARD_MEMBER(x)

have exactly the same meaning as before. When we **instantiate** the variable with an actual value, we can drop the quantifier symbols and leave a true statement about the instantiated value:

ON_STRENGTH(Smith_A_E) → PAY_NI(Smith_A_E)

means that if Smith is on-strength he must be paying NI. While

ON_STRENGTH(Smith_A_E) & BOARD_MEMBER(Smith_A_E)

means that Smith is on-strength and is a Board member.

Note that we know Smith pays NI only if it turns out to be true that he is on-strength, whereas the second formula tells us flatly that Smith is both on-strength and a member of the Board.

Finally in this introductory section, let us look at an example using a non-singularly predicated proposition. We shall use as our sample proposition: "All Board members earn salaries in excess of £30 000."

∀x. ∀y. BOARD_MEMBER(x)
→ (SALARY(x, y) & GREATER_THAN(y, 30000))

This introduces two variables into the same formula. We have the propositions SALARY(x, y) which is to be read as 'x has salary y' and GREATER_THAN(y, n) which is to be read as 'the number y is of greater numerical value than the number n'.

Clearly, x could be given the more explicit variable name 'employee', but what would we call y? Although it is playing the role of actual salary value on its first appearance, its second appearance is just 'as a number', the

GREATER_THAN proposition being typical of built-in functional predicates we met in Chapter 5. This helps to show that it is sometimes better to use the shorter symbolic name for a variable. This is in any case good for length-saving in formulae, because variables make repeated appearances in any given formula—once following the quantifiers and at least twice in the body of the formula.

Let us now agree to confirm the following typographic conventions:

- Proposition names to be wholly in upper case.
- Variable names to be wholly in lower case and as a comma-separated list where appropriate (ie. a modification of Chapter 5 conventions).
- Instantiated values of variables to be wholly in lower case except for the initial character, which will be in upper case.

If this seems clumsy just bear in mind that, while we want our modelling to be as readable as possible, it is essential to be able to distinguish these three categories of name quickly. We have to issue the usual warning that there is no universally agreed set of conventions for this.

Apart from these conventions we can adopt the relational and logic notations as appropriate.

8.1.4 Modelling and interpretation

As we shall see in the following sections, the predicate logic formulae do encourage us to believe that we have devised a means of automating reasoning in a way that bears some resemblance to our human way of thinking—at least in a Sherlock Holmes kind of way. We want to consider, though, two features which are very important to the measure of success of this activity—one a cautionary note about its limitations and the second a positive point about the power of the method.

The examples set out in the previous section are, as is very often the case in formal, mathematical modelling, something of an attempt by the writer to deceive the reader. The writer has set out a formulation of some modelling problem and says to the reader, either explicitly or implicitly, "You can see that this is right." The conscientious reader will check, the less conscientious will accept it at face value in order to get on with the text. The deception lies in the presentation of a huge conceptual leap as if it is a small but obvious step.

The formulation of our examples related to employees is, in truth, a shoe-horning operation. We have a method to ply so let us think up credible examples to squeeze into it. If this helps to make clearer what the method is—in this case the method of modelling with predicate logic notation—fair enough, but there ought to be the strongest health warning that working in the other direction, from problem to model, is a very different mental exercise. It may be necessary to knock ideas around for a considerable time and between a number of people before an agreed and acceptable model is established. How do we know that 'employees' is an appropriate variable? Do we really know what we are modelling when we formulate ON-STRENGTH as a proposition? How do we know that a user of our model will interpret things in the same way? And of crucial importance, how will we recognise errors committed by software using our models?

We are not in this text attempting to enquire into problems of knowledge acquisition and it is only fair to say that some of the above questions may be partially answered by looking into the techniques associated with that activity. But the extent of the difficulties should not be underestimated. An experienced and aware system analyst will be familiar with the burgeoning range of CASE tools becoming available to help with the modelling of information systems. This growth is a recognition that modelling data and the processes associated with data is very far from being straightforward. The early days of data processing spawned simplistic file systems that worked from the programmers' point of view but often left the business users baffled, frustrated and, at worst, without a truly effective information system.

The systems analyst should therefore keep a wary eye on the application of expert systems to business problems where that application has been effected without considerable justification for the modelling and design decisions. One of the aims of this text is to give systems analysts some confidence to take a suitable critically aware viewpoint in this development area.

The second, contrastingly more optimistic, point to be made is that, if the difficulties can be overcome, the potential power of such logic modelling is great and incorporates a kind of bootstrapping effectiveness that is not immediately obvious. As a simple example of this consider the function GREATER_THAN used in the previous section. Once such a predicate function has been formulated it is possible to start to develop theorems about it. For example, an obvious theorem for inequality is that which says

If $x > y$ and $y > z$ then $x > z$

We might model this in predicate logic form as

$$\forall x. \ \forall y. \ \forall z. \ (\text{GREATER\_THAN}(x, y) \ \& \ \text{GREATER\_THAN}(y, z))$$
$$\rightarrow \text{GREATER\_THAN}(x, z)$$

But the interpretation of x, y and z as pure numbers is only one interpretation. A closely allied interpretation is that of ages in years and of GREATER_THAN as meaning 'older than'. A less directly related interpretation is that of alphabetic sequence. There might be other specific interpretations in the real world that also fit, such as 'on a higher shelf than'. What we may realise is that the pure numbers interpretation is a special instance of a ranking relationship between two items. Theorems developed for the more general interpretation may be capable of being applied 'down' through the family of interpretations of which it is the parent. An intelligent robot may therefore be able to use this higher gear of reasoning if appropriately programmed to do so.

That last condition reminds us once again of the warning about interpretation. So, while we may be able to get a hint of the great potential power of the method, there must be the usual caveat about the use of technology being equally capable of leading us towards dangerously powerful errors.

8.2 Working with predicate logic

8.2.1 Eliminating the quantifiers

Can the universal quantifier ∀ be regarded as redundant if when writing out logical formulae it is understood that it applies to all values in the relevant domains? The logician has a neat device for showing that this is effectively the case. Let us look again at

$$\forall \text{ employee. ON\_STRENGTH(employee)} \rightarrow \text{PAY\_NI(employee)}$$

If we select arbitrarily an employee who is in the domain—let us call him 'arbit'—then ∀ has to be dropped anyway:

$$\text{ON\_STRENGTH(arbit)} \rightarrow \text{PAY\_NI(arbit)}$$

But arbit is anyone in the domain, so we could say we are going to represent such a value by a, giving

$$\text{ON\_STRENGTH}(a) \rightarrow \text{PAY\_NI}(a)$$

And we are now looking at the rule without the need for ∀! We just need to know that a is the arbitrarily selected employee.
Similarly for

$$\exists \text{ employee. ON\_STRENGTH(employee) \& BOARD\_MEMBER(employee)}$$

we can think of the one or more employees who satisfy this condition, use some symbol—q is the favourite for this role—to denote a particular one of these, and thus be able to assert that the following must be true:

$$\text{ON\_STRENGTH}(q) \text{ \& BOARD\_MEMBER}(q)$$

And again we have lost the existential quantifier. The quantifiers would be a great nuisance in any software machine implementation of the logic so it is good that they can be dropped, though the implementation must be able to guarantee that its a's and q's are from within the sets that we have explained—the domain set for a and the particular set for q.

We have only shown that they can be dropped in the simple types of example above, but it can be shown that they are removable from more complex formulae, though the manipulation can be quite difficult, especially with quantifiers embedded within nested expressions.

8.2.2 Instantiation

We have already seen that the variables in a formula can be instantiated when we know that we have a value of the variable for which the formula is true, such as Smith A.E. in our earlier example. But instantiation can be the result of deductive argument also, and this is an important advance in the quest to automate reasoning.

Let us assume that the following are given, ie. each formula is TRUE:

$\text{ON\_STRENGTH}(x) \rightarrow \text{PAY\_NI}(x)$
$\text{BOARD\_MEMBER}(x) \rightarrow \text{ON\_STRENGTH}(x)$
$\text{BOARD\_MEMBER}(x) \rightarrow (\text{SALARY}(x, y) \ \& \ \text{GREATER\_THAN}(y, 30000))$
$\text{BOARD\_MEMBER}(\text{Smith\_A.E.})$

The final line means that Smith is, as a matter of plain fact, a member of the Board. From the second line we can deduce by instantiation and mod pons that Smith is on-strength as follows:

$\text{BOARD\_MEMBER}(\text{Smith\_A.E.})$
$\text{BOARD\_MEMBER}(\text{Smith\_A.E.}) \rightarrow \text{ON\_STRENGTH}(\text{Smith\_A.E.})$
$\text{ON\_STRENGTH}(\text{Smith\_A.E.})$

And similarly we can use the first premise to go on and deduce

$\text{PAY\_NI}(\text{Smith\_A.E.})$

that Smith pays National Insurance contributions. This **forward chaining** of the argument is apparently rather aimless, but if we had put the question "Does Smith pay National Insurance contributions?" then the whole thing looks much more focussed. What we would actually do is to give to the inference engine the assertion PAY_NI(Smith_A.E.) as a goal to prove as being TRUE. Whether the engine fires all possible deductions in a forward chain until it finds the assertion, or uses the law of contradiction by assuming the assertion FALSE, adding it as a premise and then looking for a contradiction, we don't need to be too worried about. (Though if the discussions in Chapter 7 have been absorbed those options will be understandable.) What we have seen is the possibility of putting a query to the system which is triggered into deductive argument and is ultimately able to answer the question.

We have cracked an important problem—that of understanding that this basic mechanism is not only possible, but how it is possible. Anything after this is to do with putting more complex queries to inference engines which have access to ever more voluminous 'givens' in its knowledge base, but the principle of operation will be the same.

For example, we might think of putting to our simple knowledge base such questions as

> Are there any Board Members?   (Answer is Yes)
> How many Board Members are there?   (One)
> Please list all Board Members.   (Smith_A.E.)
> Please list all employees.   (Smith_A.E.)
> Does Smith_A.E. earn more than £10000?   (No)
> How much does Smith_A.E. earn?   (???)

This final question is interesting, because we have asked a question that implies a proposition not even in the knowledge base. If it were, it might be of the form:

> EARNS(employee, annual_salary)

where we have given the variables clear names. If that line was amongst the given, then the system might respond more meaningfully with something like 'Insufficient data' because it recognises now that the question:

> EARNS(Smith_A.E., x)

is legitimate, but is unable to find a fully instantiated version of it. This might have been

> EARNS(Smith_A.E., £40000)

Had this been present the answer could have been given, of course.

If we think of the more realistic situation where the full set of employees would be defined by any instantiations of a proposition type, say

> ON_STRENGTH(x)

we can see that answering a question such as 'Is Andrew_B.R. on strength' or responding to a request to 'List all employees who are on strength' would mean triggering processes which are to all intents and purposes file processing activities. A file of data in the predicate mode is indeed a list of many instantiations. But a list of all employees who pay National Insurance contributions, although by implication the same as the on-strength list, would appear to need to be generated. Whether the inference engine is capable of taking the short cut or not—recognising that it already has the list of names in this case—is a question of operational efficiency which we shall not look into here. Such questions are of course of crucial importance in evaluating the efficiency of actual software.

8.2.3   Arguments in predicate logic

We have seen then that we can apply what we established for simple propositions more or less directly to the predicate form, except that there needs to be a careful handling of the instantiated propositions in the predicate form. Let us work through some of the rules and forms established in Chapter 7.

For example, take the simpler version of the resolution rule which applies as follows with simple propositions:

A ∨ B   (given)
˜B   (given)
A   (by resolution)

Supposing A and B are predicated on the same variable x, but that what is given is an instantiated value for ˜B. We can think of this value as being arbitrary, though specific, and we shall as before call it a. The argument now goes like this:

A(x) ∨ B(x)   (given)
˜B(a)   (given)

A(a) ∨ B(a)   (instantiation)
A(a)   (resolution)

Consider this as an actual example as follows:

EXAMPLE Employees must be either Union Members or Staff Association members (or both). Johnson E. is not a Union member. He must therefore be a Staff Association member. (This example is easy enough to work out using basic common sense.)

Let UNION(employee) and STAFF(employee) represent that an employee is a Union and Staff Association member respectively. We have

UNION(employee) ∨ STAFF(employee)   (given)
NOT Union(Johnson_E)   (given)

UNION(Johnson_E) ∨ STAFF(Johnson_E)   (instantiation)
STAFF(Johnson_E)   (resolution)

A more complex example, though using precisely the same rules, occurs when we have some propositions multi-predicated. Consider the following six given formulae:

A(y) ∨ ˜B(x) ∨ ˜C(x, y)	(1)
A(z) ∨ ˜D(x) ∨ ˜C(z, x)	(2)
B(a)	(3)
C(a, b)	(4)
C(c, a)	(5)
D(a)	(6)

The symbols x, y and z represent variables and a, b and c represent actual values. We are asked to prove that A(b) and A(c) are both (separately) TRUE.

Again we shall first work this through 'mechanically' and then cite an actual example against which understanding of the reasoning may be tested.

Each line below is a combined instantiation and resolution, using the formula lines indicated:

$A(y) \vee \tilde{}C(a, y)$ (1) and (3)   (7)
$A(z) \vee \tilde{}C(z, a)$ (2) and (6)   (8)
$A(b)$ (4) and (7)
$A(c)$ (5) and (8)

Notice how variable symbols are important only as labels within a formula. In lines (1) and (2) the variable symbol x plays different roles but serves a similar purpose.

Looking at the following actual example should help to sort out the potential confusion caused by reading these symbols.

EXAMPLE   A certain circulation list includes names of personnel who act as nominated deputies to a Board Member. The list also includes names of personnel who have a Head of Department as nominated deputy. Andrews is both a Board Member and Head of Department and has Bridges as his nominated deputy. Collins has Andrews as his nominated deputy. Show that Bridges and Collins are both on this circulation list.

(Note: The reader might like to think this example through using common sense initially to see how relatively straightforward a piece of reasoning it requires.)

Let us represent this information with the following predicated propositions:

     BOARD_MEMBER(employee)
     DEPT_HEAD(employee)
     DEPUTY(employee, employee)   (The second named is the deputy)
     LIST(employee)

We can now express what is given as follows:

BOARD_MEMBER(emp$-x$) & DEPUTY(emp$-x$, emp$-y$) $\rightarrow$ LIST(emp$-y$)
DEPT_HEAD(emp$-x$) & DEPUTY(emp$-z$, emp$-x$) $\rightarrow$ LIST(emp$-z$)
BOARD_MEMBER(Andrews)
DEPT_HEAD(Andrews)
DEPUTY(Andrews, Bridges)
DEPUTY(Collins, Andrews)

To aid cross-comparison, we have used personnel whose names begin with corresponding characters and suffices on the employee variables to match the variable symbols used in the formal example.

The first two formulae can be rewritten to match (1) and (2) above by expressing in normal form as follows:

LIST(emp−y) ∨ ~BOARD_MEMBER(emp−x) ∨ ~DEPUTY(emp−x, emp−y)
LIST(emp−z) ∨ ~DEPT_HEAD(emp−x) ∨ ~DEPUTY(emp−z, emp−x)

It should now be seen that the argument can proceed exactly as before, but with the more descriptive symbolisation, to arrive at the required results:

LIST(Bridges)
LIST(Collins)

8.3   Some practical considerations

8.3.1   Computational sequences

As we have said repeatedly throughout this text the intention is to explain principles rather than to go into specialist knowledge about the software science and engineering techniques that are absolutely vital to the successful implementation of the concepts. We are once again at a point where the boundary is relevant. We might appreciate how a software inference engine may in principle process a set of 'givens' in order to come up with proven 'answers', but even our simple examples should have made clear that our own working through has had an arbitrariness about it.

When faced with complex logic expressions the reduction to Horn clause form ensures simplicity of the right-hand side, but the left-hand side might well remain as a lengthy conjunction of sub-clauses. Also in any problem of significance there is likely to be a large number of 'givens', many of which are general variable-using formulae, and many instantiated formulae which can be used to instantiate these experimentally. The potential number of options as to computational sequence can quickly become so enormous that even powerful processors could be operationally tied up for unacceptable periods of time.

This brings us back to the discussion about search strategies which was touched upon in the final section of Chapter 6. The blind-search approach will fairly obviously be that which takes the list of 'givens' (premises) as it finds it. If not in Horn clause or other appropriate clausal form already, it will process all premises until such standard forms are achieved—resulting in a longer list of course. It is reasonable to expect that basic outsorting will need to take place, such as separating general from instantiated formulae, rearranging clauses internally according to occurrence of negative atoms, and ranking formulae in order of number of clauses and negativity of atoms. These are heuristic actions, reflecting in part a common-sense approach to preparing any material for processing. The careful identifying and placing of negative atoms is in fact a very important heuristic since it anticipates the attempted matching of complementary atoms in the successive application of the resolution rule.

The actual sequence of firing of rules depends upon a number of factors. A depth-first search implies that formulae are exhaustively tested by left-to-right processing of sub-clauses within the now well-ordered formulae. But of course each firing itself generates another potential result to be used for subsequent firings and further heuristics are required to decide where these should be placed on the priority list. Such heuristics not surprisingly can in part be determined by the application context. Knowledge processing in any particular contextual area will have local characteristics whereby some processing patterns may be more common than others and therefore certain clues and signals will be appropriate. The term *demons* is often used to denote sub-programs whose specific task it is to scan a knowledge base scouting for particular types of formulae which may, pragmatically, lead to a goal more quickly. This scanning includes of course looking at the semantic (meaning) context of the propositions.

The chaining sequence of successive rules of inference is, as we have mentioned already, referred to as forward chaining if the start point is the set of premises and the processing is directed towards the solution goal or goals. If the start point includes the formula or formulae created by incorporating the goal (or more likely its negation in using the law of contradiction) and trying to process towards tautology or contradiction, this is referred to as **backward chaining**.

Sequence in computation also implies that some action is taken, results stored and control taken on to the next action. Is the step reversible? By some kind of total recording and archiving it must always be theoretically possible to wind back and undo the logic processes, but there are implications here again of prohibitively high cost once the number of steps gets to be more than a few—or even if it is just one where that step has in effect involved the processing of a large file of formulae. However, such *backtracking* has to be available to the program which is engaged in processing the formulae since it must be able to leave a formula intact if it finds that after search of all clauses there is no firing possible. This is not too much of a problem since the search algorithms have been designed and optimised to cope with this.

But where we are thinking of the need to reverse the firing action of rules the problem is a much greater one since actions will in general have altered actual contents of the knowledge base. This is known as the problem of **monotonicity**—a term which unfortunately does not convey well to the non-specialist that it is referring to direction and irreversibility of reasoning. Non-monotonic systems are rare, except in the sense that one step may be undone as is quite common with spreadsheets and expert system shells, or in the less-acceptable sense that the user has to do the work by means of back-up copies for example.

We have approached predicate logic from the direction of logic rather than from the direction of natural language because this text has aimed to introduce principles of logic first. However claims are often made that predicate logic is promisingly close to natural language and that users of knowledge processing software systems may soon be able to use more natural interface languages based on predicate logic, without having to understand much at all about the principles of formal logic. Whether these claims are justified or not is still in some doubt but it is certainly worth looking at some of the possibilities and difficulties involved when attempts are made to turn the logic language into more natural forms. (The issue of devising forms for natural language which satisfy logical rigour is a rather different one. We shall turn to it briefly in the next chapter.)

By citing the proposition name followed by, in brackets, the predicate items, we have so far been using what is called the **prefix notation** for predicate logic. If the predicate items are distributed around a phrase-framework type of proposition so that the whole formula reads more like a natural language sentence, then this is referred to as an **infix notation**. The principles are exactly the same—this is purely a setting-down device in order to create more readable-looking results. There is an inevitable penalty of greater overall length of formulae but where the intention is that ordinary users should be able to use the software, such as in general education or by general management staff, there is an obvious attraction to this approach.

Some means of being able to distinguish between proposition names, variable names and actual values must be retained however, so completely natural-looking sentences seem to be ruled out for the time being. Also positioning of elements within sentences must be rigidly consistent and this is far from natural if we look at most written narrative and uttered speech.

Let us take an example from an earlier section and transform it from prefix to infix form. For example,

$$\forall \text{ employee. ON\_STRENGTH(employee)} \rightarrow \text{PAY\_NI(employee)}$$

could be re-written as

> For all employees who are on-strength it is implied that these employees pay NI.

A crude analysis of this sentence for correspondence with the logic might be

> For all employees (who are) ON-STRENGTH (it is) implied that (these) employees PAY NI.

Underlined words are the translations of the operational symbols and the words in brackets indicate what we have had to add because of the lack of good grammar and style in the prefix format.

Let us accept that 'For all' could be taken for granted if we recognise 'employees' as a variable name. Also a compromise in style between the two formats might give us the advantage of retaining the consistency of the prefix format while gaining some naturalness from the infix format. Let us try therefore:

employees ON-STRENGTH implies employees PAY NI

This would retain whatever naturalness it has even if we resort to x for the variable name:

x ON-STRENGTH implies x PAY NI

Whether this is accepted as still being within the category of 'readable' will be open to debate, but we hope that the flavour of the nature of the problem is beginning to come across.

Let us try another example:

∃ employee. ON_STRENGTH(employee) & BOARD_MEMBER(employee)

could be re-written:

There is at least one employee who is both on-strength and a Board Member.

We can analyse this as

There is (at least one) employee (who is) both ON-STRENGTH and (a) BOARD MEMBER.

The compromise is more difficult here because the existential quantifier is not easily taken as read if omitted and 'employee' has to be attached to both propositions connected by the logical and. We can try using the device of citing an actual, though anonymous, employee (call it A-employee) who we know must exist by virtue of the existential quantifier:

A-employee ON-STRENGTH and A-employee BOARD MEMBER

Having to recognise that A-employee means 'actual though anonymous' is hardly like natural language at all, but it does raise awareness that natural language interfaces are not going to be easy to develop even in a relatively well-defined context such as provided by predicate logic. However, the example is a little more readable when we cite an actual person since we can now resort to listing the and linked atoms as separate facts:

Smith A.E. ON-STRENGTH
Smith A.E. BOARD MEMBER

An instantiated simple implication such as

ON_STRENGTH(Smith_A_E) → PAY_NI(Smith_A_E)

translates fairly straightforwardly as

> Smith A.E. ON-STRENGTH _implies_ Smith A.E. PAY NI

Notice, though, how we may have to live with the poor grammar of 'Smith pay NI'.

With double-predicate examples the proposition name can be infixed between the two predicate names with reasonable comfort. Consider

> $\forall x.\ \forall y.\ \text{BOARD\_MEMBER}(x)$
> $\rightarrow (\text{SALARY}(x, y)\ \&\ \text{GREATER\_THAN}(y, 30000))$

This might translate to

> x BOARD MEMBER _implies_ x HAS SALARY y _and_ y GREATER THAN 30000

HAS SALARY reads better than just SALARY in this infix form so the logic version is asked to make the sacrifice. But although in natural language we would be almost certain to say 'where y is' rather than 'and y is', it is too much to ask the translation to allow this change. (The 'too much' refers to our current aim of seeking an easily implementable compromise.)

Notice that even when we instantiate this rule for an actual individual, the salary amount stays as variable:

> Smith A.E. BOARD MEMBER _implies_ Smith A.E. HAS SALARY y _and_ y GREATER THAN 30000

We could of course use some general variable name like 'amount' instead of y to increase readability:

> Smith A.E. BOARD MEMBER _implies_ Smith A.E. HAS SALARY amount _and_ amount GREATER THAN 30000

The name 'amount' applies only to this sentence though. It could be used in the very next sentence and mean something quite different. For example,

> stock-pack VOLUME amount _and_ amount GREATER THAN 25 _implies_ stock-pack WAREHOUSED B-Shed

The sudden change of subject from Personnel to Stock Control may be very unlikely in practice of course. The aim was to make clear the very local role of variable names in a language based so closely on predicate logic.

Finally, to make a translation from a triple-predicate (or higher-order) proposition requires creation of additional names and structure to the sentence form. Let us take the example:

> $\text{RUNS}(o, m)\ \&\ \text{WORKS\_UNDER}(p, o) \rightarrow \text{COMPATIBLE}(p, m, o)$

which means, say,

> If a certain operating system runs on a given machine and a certain

programming language works under that operating system, then the language is compatible with that machine for that operating system.

(An experienced programmer may want to dispute the wisdom of this rule!) The translated form of this might look something like:

operating-system RUNS ON machine <u>and</u> programming-language WORKS UNDER operating-system <u>implies</u> programming-language COMPATIBLE machine FOR operating-system.

The change from RUNS to RUNS ON is purely for style, but the change from compatible to COMPATIBLE...FOR... is a necessary consequence of infixing. The system would have to remember that the phrase is monolithic and must have three variable names or instantiations associated with it in the positions where there are dots.

From the natural language point of view all of the above constraints of name-control, word-position and stripping away of redundant words still verge on the disastrous. The logician may be proud of the progress made in making the obscurity of symbolic logic look more like natural language, but when it comes to actual use of such languages there is much more the flavour of programming-with-phrases than with reading, writing or speaking. And experienced interface users may soon grow tired of the tedium of excessive word and phrase length and seek to return to the parsimony of symbolic forms. But it is a field currently receiving considerable research attention and we shall return to some of these issues in the next chapter.

8.4 Prolog—a language for predicate logic

8.4.1 The language

A famous first in the world of logic was the launching of a language for PROgramming in LOGic (hence Prolog) in the early 1970s by Alain Colmerauer and his research team at the University of Marseilles. The design of this programming language was firmly based upon the calculus of predicate logic. Its appearance caused great excitement and its potential and promise soon led to its being taken up as an important new tool for artificial intelligence work. The announcement by Japan in 1983 that Prolog had been selected as the system language for their fifth-generation computer development project was taken as an indication that they really meant business and were not simply playing with high-blown theory.

While the actual achieved development in the Japan project was looking disappointing by the end of the Eighties, the fright to Europe and USA had been enough to generate enormous research and development effort in artificial intelligence (and specifically the Alvey Initiative in the UK), and not least among the areas receiving direct attention was, not surprisingly, Prolog itself.

As an implemented language Prolog has appeared in several incarnations, so there is the inevitable problem of lack of standard format and even of standard definition. But in one incarnation or another it has been used in many real and important applications, not the least being for the development of expert systems, either directly or as the language in which expert system shells have themselves been written. True, as experience has shown that software efficiency is difficult to attain with Prolog (some Prolog-based systems have later been re-written in C for example), the bloom has gone off the plum a little. But the stimulus provided by Prolog being in the right place at the right time can never be gainsaid and, in any case, there is almost certainly still a bright future for the language as a powerful development tool.

And not simply in the field of artificial intelligence either. There are reasons to believe that it can fulfill the long-promised role of presenting predicate logic in a reasonably natural language garb, its uses ranging from the quite humble ones of everyday handling of information to the more sophisticated ones which have already included real-time industrial control systems, language translators and theorem proving. Although its relative lack of operational efficiency may prevent its becoming a standard tool for large relational database development and use, it has attractions as a tool in this respect for the personal computer user.

With this last point in mind, we are going to look at Prolog through a particular implementation called Turbo Prolog developed by Borland International and first released in 1986. Prolog purists have criticised Turbo Prolog for being less than the full language specification, but it has the merit that it was specifically developed for the average PC user as an inexpensive, easy-to-learn yet patently useful version of the language. It also has more than enough of the range of features to enable us to use it here as a vehicle for quickly learning what a logic programming language can look like.

8.4.2   The main features of the Prolog language

The basic format of a Prolog sentence is of the general Horn clause shape:

    A if B and C and ...

The consequent is thus written first before the if, and if thus represents the leftward pointing arrow, as in

    A ← B and C and ...

The and is the conjunctive connective (AND, &, etc.). Strictly speaking no other connective is required, since negation (NOT, ~, etc.) can always be incorporated by a renaming substitution (Y for ~X) and disjunction (OR,

V, etc.) can always be eliminated by Horn clause generation as discussed earlier. However, either or both of these may be made available in actual implementations as a convenience to the user. Predication is displayed as the proposition name followed by a comma-separated list of variable names within parentheses.

The typographic conventions vary across the different implementations, but those for Turbo Prolog may be illustrated by a typical Turbo Prolog sentence (taken from a murder solving example!):

suspect(X) if owns_probably(X, Object) and
    same_wounding(Object, Weapon) and
    killed_with(susan_shaw, Weapon)

The first thing to note is that proposition names (eg. suspect, killed_with, owns_probably), connective symbols (eg. if) and instantiated value names (eg. susan) are all fully in lower case. The Turbo Prolog interpreting program relies upon brackets, commas and space characters to identify and discriminate between these. Variables are recognised by their initial or only character being in upper case (eg. X, Weapon, Object).

A possible reading of the above sentence is therefore

Mark as a suspect any person who there is reason to believe probably owns an object which could cause the same wounds as the weapon which was used to kill Susan Shaw.

This is still not as natural as the way a detective might actually express this piece of reasoning in spoken or written instructions to colleagues and assistants, but we hope it is natural enough to be understood as a piece of narrative while at the same time being capable of conceptual referencing across to the Prolog sentence. Apart from the obvious typographic and vocabulary changes, notice the changes in structure leading to the use of 'which' clauses to cope with the logical variables.

Infix notation might look more natural than this Turbo Prolog format, but the main issue being pointed to here is the need to be able to identify the logical structure in English sentences in order to convert them into the Prolog form at all. How clear is it that the consequence (the Horn clause head) is the marking as suspect of an individual? Or that object ownership by an individual and object likeness to the murder weapon are totally variable-handling propositions? Not very. However we have here started in the middle of the story, which partly (but only partly) explains such difficulties.

Let us go back to beginnings. A Turbo Prolog program comprises a number of sections. We shall be interested in four of these, called domains, predicates, clauses and goal. The vocabulary we have developed through the text is of help here but we do need to look at the precise meaning of these section names rather carefully.

8.4.3 Domains and predicates

The domains section is merely a listing of names and types of data we want to use in the predicated propositions, called simply predicates in Turbo Prolog. The predicates section is thus a listing of these predicates with the appropriate data type names from the domains section. Thus these two sections merely set out the basic material and do not contain any logic as such. Our murder example might have the following domains and predicates sections:

```
domains
    forensic, murder_weapon, occupation, person, sex, weapon = symbol
    age = integer
predicates
    had_affair_with(person, person)
    interview_report(person, age, sex, occupation)
    killed_with(person, murder_weapon)
    owns(person, weapon)
    probably_owns(person_weapon)
    same_wounding(weapon, murder_weapon)
    smeared_with(person, forensic)
    suspect(person)
```

What we have done here is to specify names of predicates that we intend to use in later sections and we also indicate the type of data each of these predicates will use by specifying what are called object names (eg. person, weapon, etc.) within the brackets. Note that these are neither variables nor instantiated values, though like the latter they comprise lower case characters only. Prior to this, these object names are used to indicate that type of data they represent: char, integer, symbol, string, real are the basic alternatives here, and symbol is the most general, meaning that the data may comprise both alpha and numeric characters. (We shall not describe the rules in rigorous detail, the aim once again being merely to give here a flavour of the language.)

8.4.4 Clauses

The next section is the clauses section and it is here that actual information is provided in two ways—as facts and as rules. It helps if one regards the definitions in the predicates section as potential tables (or relations, or files), the object names being the column heads. For example it seems very likely that the interview_report predicate was set up to enable the basic data about each interviewee to be entered as facts. We might, for instance, have

interview_report(john_andrews,   24,   male,   draughtsman).
interview_report(beryl_tate,   30,   female,   nurse).
interview_report(colin_pritchard, 50,   male,   bus_driver).

and so on. Clauses must be terminated with a full stop. The spacing has been inserted to aid readability.

These 'fact' clauses contain instantiated values and they are of the Horn clause head-only type, as dicussed in Section 8.3.3. They are always TRUE, which is what we want facts to be!

Other predicates may be fed with appropriate facts too, of course. For example,

owns(john_andrews, golf_clubs).

The question now arises as to why we haven't included weapon ownership as one of the objects in interview_report. The answer is that we wish to incorporate this clause in a later rule clause. For example,

suspect(Person) if owns(Person, Weapon) and
                   same_wounding(Weapon, heavy_club).

Note the upper case first characters of the variable names here. Person and Weapon could have been called X and Y to emphasise their roles as variables, but we have gone for readability instead.

So if triggered, will the inference engine conclude that John Andrews is a suspect? No, not yet. But if the following fact clauses were to be added:

killed_with(susan_shaw, heavy_club).
same_wounding(golf_clubs, heavy_club).

then he most certainly would become a suspect. By a relentless operation of its resolution-rule machinery, the inference engine would eventually fill in the name john_andrews into the person position in the predicate suspect(person). But we said 'if triggered'. What trigger? We look at this next.

8.4.5   Goals

The goal section appears between the predicates and clauses section (though it may be omitted, when the Turbo Prolog program will prompt the user for goals after reading-in the clauses section.) The goal (or goals) is a statement by the user of some clause which invites the Turbo Prolog program to say whether it is TRUE or not. In our example we might want to check the nature of the murder weapon. So

killed_with(susan_shaw, club).

as a goal will yield the response:

True.

On the other hand,

killed_with(susan_shaw, knife).

will yield the response:

No solution.

because Turbo Prolog has been unable to resolve this with any of its facts, direct or inferred. But if we set the goal

killed_with(susan_shaw, Weapon).

we are giving the system an unknown variable to work on.
  (Perhaps

killed_with(susan_shaw, X).

looks more like an unknown, but to Turbo Prolog it is exactly the same task.)
  The task is to try to fit an instantiated value where the variable is given. This time the system responds with

Weapon = club
1 Solution.

It gives this response because it has been able to satisfy Weapon with the value club which appears in the TRUE clause:

killed_with(susan_shaw, club).

Another example, giving more than one solution, is the goal:

goal
    interview_report(Person,_,_,_).

This makes use of anonymous variables—the underline character—which means 'may take any value' rather than 'What are the values?' The response

Person = john_andrews
Person = colin_pritchard
Person = beryl_tate
3 Solutions

should by now be reasonably easy to understand. By making one of the anonymous variables specific we will again reduce the number of solutions, as with

goal
    interview_report(Person, 50, _, _).

which evokes the response:

    Person = colin_pritchard
    1 Solution

or

    goal
        interview_report(Person, _, male, _).

which yields

    Person = john_andrews
    Person = colin_pritchard
    2 Solutions

The sequence of the solutions is significant, being the order in which Turbo Prolog itself hits on the solutions. Here it is simply dependent upon the sequence of filing of the facts clauses of course. Where inference has taken place it can be much less obvious as to why the sequence is as it is.

Finally, before looking at a fuller example of this Murder Mystery program, consider a goal which does set rule inferencing into action, and at last answers the question posed at the end of the previous section. Namely, how do we get to know whom the program suspects? We do this simply by posing:

    goal
        suspect(X)

using the X unknown as a change. Assuming we have all the clauses mentioned in that previous section now safely installed under the clauses section, we shall get

    X = john_andrews
    1 Solution.

8.4.6 Other features

We may by now be beginning to think that this is such a pathetically simple piece of detection that it hardly seems worth the bother! We have been keeping it simple to help illustrate the principles involved of course. Let us now look at a fuller version of the program and use it to illustrate some other features not yet discussed. The program is set out as Fig 8.1. In this program we have used the comments feature—any text enclosed between /* and */—which helps to explain the intention of the program writer. The Turbo Prolog compiler ignores such text.

The story appears to be as follows (try to trace the appropriate clauses in Fig 8.1):

It is Susan Shaw who has been murdered, having been bludgeoned

/*Murder Mystery*/

domains
    forensic,murder_weapon,occupation,person,sex,weapon = symbol
    age = integer

predicates
    eliminated(person)
    had_affair_with(person,person)
    interview_report(person,age,sex,occupation)
    killed_with(person,murder_weapon)
    motive(person)
    owns(person,weapon)
    probably_owns(person,weapon)
    same_wounding(weapon,murder_weapon)
    smeared_with(person,forensic)
    suspect(person)

goal
    write("Primary suspects")   /*Report head*/
    suspect(Person)

clauses

/*The case*/
    killed_with(susan_shaw,heavy_club).

/*Those interviewed so far*/
    interview_report(john_andrews,   24, male,   draughtsman).
    interview_report(beryl_tate,   30, female, nurse).
    interview_report(colin_pritchard,   50, male,   bus_driver).
    interview_report(phillip_turner,   22, male,   teacher).
    interview_report(jean_hadleigh,   43, female, teacher).

/*Some are already eliminated from our enquiries*/
    eliminated(jean_hadleigh).
    eliminated(Person) if interview_report(Person, Age,_,_) and Age > 45.

/*Relevant recent sexual liaisons*/
    had_affair(john_andrews,   susan_shaw).
    had_affair(john_andrews,   beryl_tate).
    had_affair(phillip_turner, beryl_tate).

    had_affair(X,Y) if had_affair(Y,X).

    motive(Person) if had_affair(Person,susan_shaw).
    motive(Person) if had_affair(Person,X) and
                      had_affair(X,susan_shaw).
    motive(Person) if had_affair(Person,X) and
                      had_affair(X,Y) and
                      had_affair(Y,susan_shaw).

/*Those with access to possible murder weapons*/
    owns(john_andrews,   golf_clubs).
    owns(beryl_tate,   scissors).
    owns(phillip_turner, cricket_bat).
    owns(jean_hadleigh, sword).

    same_wounding(golf_clubs,   heavy_club).
    same_wounding(cricket_bat,   heavy_club).
    same_wounding(large_spanner, heavy_club).

    probably_owns(X,Y) if owns (X,Y).
    probably_owns(Person,large_spanner) if
                interview_report(Person,_,_,bus_driver).

/*Forensic evidence*/
    smeared_with(phillip_turner,   lipstick).
    smeared_with(colin_pritchard, blood).

/*Generate suspects list*/

  /*on basis of access to weapon*/
  suspect(Person) if probably_owns(Person,Weapon) and
                same_wounding(Weapon,Murder_weapon) and
                killed_with(susan_shaw,Murder_weapon) and
                not eliminated(Person).

  /*on basis of forensic evidence*/
  suspect(Person) if smeared_with(Person,_) and
                not eliminated(Person).

  /*on basis of motive*/
  suspect(Person) if motive(Person) and
                not eliminated(Person).

Fig 8.1 The Prolog murder-case knowledge base

with a heavy club of some kind. Jean Hadleigh has been eliminated from the enquiry for some reason not given and it has been established that the murderer could not have been older than 45 years of age.

Golf clubs, cricket bats and large spanners are all possible murder weapons in this respect and in any case among the interviewees ownerships of all potential weapons have been recorded. It is felt that an occupation such as bus driving would imply access to a large spanner. Traces of the victim's lipstick have been found on Phillip Turner's clothing and traces of her blood on Colin Pritchard's.

There have been three affairs involving three of the five people who were interviewed. An affair should be taken as a linking in either direction between the two involved. Anyone who has been involved in sexual liaison with the victim or involved at one or two removes through intermediary partners is regarded as having a motive for the murder.

Finally we are placing on the suspect list any person not eliminated from the enquiry for whatever reason who owns or has access to an appropriate weapon, who recorded positively on forensic evidence or who has been adjudged to have a motive for the murder.

The suspect list which the program will produce when run will comprise:

John Andrews
Beryl Tate
Phillip Turner
John Andrews
Phillip Turner
Phillip Turner.

The first three names come from the first suspect(Person) clause, the next two from the second and the final name from the third. This is why there are repeats. In some uses such repeats are a nuisance and there is a way to avoid them which we shan't discuss here. The weight of the case against Turner could be made more directly obvious if we had set up a single suspect(Person) clause and connected all the different component clauses of each with and. His name would have then been the only one on the solution list! This is because his is the only name that satisfies all the components, which is why of course his name comes up three times in the present version of the program.

Some features not yet discussed appear in this program. In order of appearance, they are

The write clause in the goal section. This outputs to the screen the text enclosed between double quotes. It can also be used to write selectively from the results of a goal search, overriding the crude listing of all the hits, which is the default action.

The arithmetic comparator clauses, using an infix notation. Here we have forced into the eliminated predicate all interviewees whose age is greater than 45. In Turbo Prolog, as in most computer languages, there are a number of such relational operators for placing conditional constraints upon numerical data values.

The not connective. Logical negation has been used to ensure that those whose names are in the eliminated predicate do not get included in the suspect lists.

There are many other features which help to give Turbo Prolog great power and usability. We shall only mention them here in passing. The reader should turn to the product itself to gain a proper understanding of the features.

Useful operational features include:

A multi-window based interface.
Screen-editing facilities.
Debugging and program tracing for fault diagnosis
Control over the solution search process.
Handling of files and input/output processes.

Features of the language itself which greatly extend programming power include:

Compound objects and lists
Any object can itself be defined as a predicated object. For example, our interview_report might need to have incorporated all the dates when interviews have taken place. We would include in the predicates section declaration an additional object dates:

interview_report(person, age, sex, occupation, dates)

But dates itself has to be signalled as being a structure rather than a simple data type. This would be effected in the domains section as follows:

dates = dates(yymmdd)
yymmdd = integer.

The name yymmdd has been used to help remember the format for the dates. For example, 900919 means here 19 September 1990. The actual dates for a particular interviewee are then entered directly in the appropriate fact clause:

interview_report(john_andrews, 24, male, draughtsman,
dates(900610, 900612, 900615)).

which records that Andrews was interviewed on three closely successive days in June 1990.

If the special compound object format for lists is used (not discussed here), Turbo Prolog has a set of commands which effectively turn it into a list processing language of some versatility. Users can, for example, get involved in the wonders of recursion. Also compound objects and lists can be taken to many layers to create tree structures.

Graphics and sound
Control of colour and dot and line manipulation enable the adept user to produce graphs, pictograms and even animation. There is also built into Turbo Prolog a set of commands called Turtle Graphics. These control the movements of a screen turtle with a 'pen attached to its tail'. Drawing is then a consequence of driving the turtle in various directions with the special commands. Sound can be controlled in terms of frequency and time-length of note and appropriate notes then associated with keyboard keys.

All-in-all Turbo Prolog provides the average personal computer user with an AI tool which, although calling for a certain amount of ingenuity from the user, does bring early rewards to the user who is willing to invest a little time in learning it.

9 Knowledge Representation

9.1 Understanding what is meant by Knowledge Representation

9.1.1 Introduction

Since knowledge is something that anyone and everybody would claim to acquire and make use of, probably without even being able to define precisely what the term means, there are a great number of possible ways of approaching the problem of its representation. However, we are here of course concerned with representation for the express purpose of some kind of automatic machine handling of the knowledge, and this, in practical terms at any rate, focuses our attention on particular aspects of what we mean by knowledge.

It certainly implies that knowledge must mean something that can be described with a degree of formality. Thus knowledge as to how to calculate an employee's monthly salary is a very promising candidate, even if we base that statement only upon the crude fact that it is a procedure used a million and more times a month around the country. Knowledge of the behaviour characteristics of one's immediate family, intimate and insightful as such knowledge may well be, seems to be a very doubtful candidate not only because of the individual characteristics of the pieces of knowledge involved but, more importantly perhaps in the present context, because we would find it very difficult to imagine how we might describe the knowledge in other than very informal natural language. Also the challenge of describing it would be less daunting if we were thinking of giving out our knowledge as a face-to-face presentation when the full gamut of natural language, body language and emotive expressiveness would be at our disposal.

The salary calculation case draws our attention to an existing store of recorded knowledge which is especially important in the context of our aims in this text. There is already in existence an enormous range and quantity of software and stored data associated with computer-based information systems. Is not this a vast treasurehouse of already represented knowledge, especially since it meets the criterion of being accessible to processing by an automatic machine? Indeed it is, and it is of great economic importance that those responsible for the use, upkeep and development of these systems should take this same attitude. The view that traditional information systems are 'old hat' and that the future belongs to expert systems and AI is a dangerously naive one—though it has to be admitted that as with all naive convictions it does contain a grain of truth.

Returning to the puzzle about what is representable knowledge and what isn't, a clue lies in our common-sense expectation that what is held statically and explicitly as a basis of knowledge is simply the raw material for the generation of a wealth of further applicable knowledge. Lengthy lists of facts printed out following accesses to a data file provides only the first step to applicable knowledge. A listed report of customer accounts may be theoretically the source for information about how to set credit levels for the next accounting period, but that knowledge about credit levels is still largely implicit. A further program may be written to sort, aggregate and select data which may then be capable of presenting that knowledge more explicitly. The issue then arises as to where can we obtain the knowledge about how to evaluate credit-limit setting methods. From the program itself? How explicitly? Surely not very explicitly at all, unless it is accompanied by some pretty remarkable documentation! And then we find we are touching once again on the problematic question as to how best to represent this qualitative kind of knowledge.

Although knowledge processing and expert systems development ideas have developed fairly independently of traditional data processing and information systems ideas, as soon as applicable systems come into actual use realisation dawns that access to the existing bases of fairly raw data is a very relevant issue. The sophistication of the expert systems is currently often impoverished through lack of access to the high data volumes of the files and databases. Had we better not then develop new systems from now on with this thought in mind? And this does not imply needing to throw away existing systems analysis design skills to be replaced wholesale by knowledge processing system design skills. This is why we have in this text unapologetically taken a systems analysis perspective on the question of knowledge representation.

9.1.2 A systems analysis perspective

There are many ways in which the occupation of systems analysis needs to develop a more robust and forward-looking view of knowledge (as distinct from mere information) processing. Here we shall discuss the more central of them—those concerned with the design process.

The formulation of design knowledge is a most urgent issue. Many tens of thousands of computer-based systems have been built over the last three decades using design methods that have ranged from being ill-defined to rigorously defined, from being informal to formal, from yielding successful to disastrous results. Nor are these easily cross-referenced scales—the ill-defined and informal design method has not necessarily yielded the unsuccessful result. The problem is that the variety of method has made the representation of the inherent design knowledge difficult to the point of infeasibility. In recent years the spread of *structured methods* has begun to offer *some* consistency of representation, but the situation is still recognised as being in need of further standardisation before we can begin to expect to be able to tap the design knowledge inherent in a developed system.

One very direct and practical consequence of this has been the disproportionately high cost of maintenance and modification. More explicit design knowledge will ease this cost. Take a fairly simple example such as needing to know, because of changes in criteria, where, how and under what conditions a comparison is made between two data items currently stored in different record types in a large system. The criteria change might, for example, be to do with the relative weighting of qualifications and experience, when it is therefore necessary to locate and analyse all comparisons made in the system between an individual's qualifications and experience and the qualifications and experience needed for promotion, placement on projects and eligibility for transfer to a sister company. Good documentation and a well-developed data dictionary system will be a great help towards planning this modification, but a knowledge handling system whereby the analyst can enquire of the design knowledge base for all the 'where, how and under what circumstances' details would be help of another order of magnitude better in quality.

This is direct use of design knowledge. What is becoming of equal importance is the *re-use of design knowledge*. To a programmer, re-usability is a question of being able to lift chunks of coding from a working system and drop them into a newly developing system with a minimum of adjustments. Ability both to spot the possibility of such re-use and also to carry out the actual operation successfully relies upon intimate knowledge of the existing code. If we step up a level or two to enquire as to the feasibility of re-using design ideas and specifications we can imagine an even more hit-and-miss affair—unless again there exists a design knowledge base together with the means of interfacing with it.

That phrase 'the means of interfacing with it' is a clue to the next point—that the knowledge has to be economically accessible. One of the early consequences of using structured systems analysis methods was the production of enormous amounts of documentation. Five hundred sheets of carefully documented data flow diagrams, data structure charts, cross-reference matrices and data dictionary record entries, even if these are electronic sheets, do not comprise useful knowledge if the potential user of it does not have ease of access to the implicit knowledge it carries. For smaller system developments there have been short-cut routes through the methodology proposed to cut down on the volume of documentation—a practical approach to the problem, but it could also be seen as running away from it in some respects.

A short-cut or simplified version of a methodology itself implies that this has now hit upon the right level of design knowledge generation for certain circumstances, but it is of course only the smallest part of the story. For what is simple for one user is not so simple for another, and is over-simplistic for yet another. The ideal is a full and sophisticated design knowledge base with flexibility in the level of interfacing with it. Ideally, the full range from most-experienced systems analyst to responsible user should be able to use the design knowledge in ways most comfortable to themselves.

And then of course there is the question of design methodology knowledge itself. Knowledge bases which provide users with how-to-use guidance and insight are also of importance if design methodology itself is to become a communicable set of design languages.

Looking even deeper into the design issue, there is the ideal of the adaptive system, the system that adapts not just to the level of sophistication of the user but which monitors and anticipates change, prompts users to question current relevance and up-to-dateness and so on. Such *cybernetic system* issues imply that design methodology knowledge also needs to be subject to knowledge processing principles. We are probably now, though, back at the difficult-to-represent level that we spoke of in the previous section, but it does help to show that the systems analysis perspective on knowledge representation is a vital one and one with, for the time being, a distant achievement horizon. However, there is plenty of visible terrain to be covered well before we close on that horizon.

9.1.3 Knowledge representation in context

By talking in terms of perspectives and by looking in particular at the systems analysis perspective we have more or less implied that the adequate representation of knowledge is relative to context. This is certainly true in the pragmatic sense—approaching a knowledge representation problem where investment in large databases is a historical fact will be a more constrained exercise than where the problem lies in virgin territory as far as computer-based application is concerned. In a more theoretical or ideal sense the picture is not so clear. Theories and continuing research have not yielded strong guidelines to date as to the question of which representation schemes are best suited to what circumstances. Indeed there is still quite fundamental disagreement among theorists as to what approaches and methods constitute sound bases for knowledge representation.

In the wider field of AI, knowledge representation is expected to be able to serve a very varied type of user. It is required to be capable of supporting descriptions of the real world, ideally in all its variety of manifestation, so that so-called intelligent machines can use it to make deductions which are capable of leading those machines to take appropriate actions. Those actions should not only be such as to perform defined routines and be adaptable to changes of definition, but also they need to be *safe* actions, first in respect of human safety, second in respect of protection of major plant, and third in respect of the continued safe-running of the machines themselves. This is the realm of robotics, cybernetics and systems science and we do no more here than pay it passing homage.

The systems analyst will usually be much more concerned with the narrower view of knowledge representation, which is the one that takes it to mean the provision of a scheme for data and rule storage in the context of some fairly well-defined application area. That area may be fairly broad—office automation is an example—or it may be quite specific—as with activities such as financial investment, production management, warehouse control, personnel management and so on. The application of expert system shells has often been in a very specific field such as fault diagnosis, income tax advice or information retrieval. But it is there that we must sound a note of caution, for one of the aims of this text is to promote the view that the systems analyst is responsible for developing systems in the wider context of business and industrial computing, and this means avoiding the pitfalls of developing systems in isolation. Which also reminds us again of the relevance to design methodology itself.

The systems analyst will usually be constrained by resource and time budgets geared to the business cycle of events. Rarely does the practical analyst get involved with research-type budgets. An essential question will therefore be that of the feasible obtainability of the knowledge material. Knowledge acquisition is a crucially important process and should be recognised as such. The analyst's fact-finding skills will serve only as the first platform to new skills and the reader is urged to make serious effort to examine this issue through other reading since this is not the text within which to present such material.

Matching the representation scheme to the problem in hand, once the acquisition exercise is at least under way, is however a relevant issue before we look at the range of such schemes, or approaches as we shall now be calling them. The following are the major aspects that need to be investigated.

Fundamental of course is that the approach be capable of expressing the detail down to the right level. An approach which treats the knowledge as uniform pieces with regular shapes to their context is clearly not suited to a problem domain where two pieces of knowledge are hardly ever the same. A Help Desk knowledge system designed to cope with a wide variety of types of question about topics that generate varying types and lengths of answer would be a case in point.

Next is the question of selection of type of inference engine. Establishing what type of engine is available may be no easy task of course, but it can generally be ascertained whether it has been designed to cope with potentially deep tree structures of inference or for relatively shallow and even perhaps fixed structures such as small pseudo decision tables. Fault diagnosis on a particular type of internal combustion engine may be fully mappable for instance, leaving the inference engine with path tracing hardly worthy of the name 'inferencing', whereas producing tactical guidance to a user of a complex system such as a power plant or a battlefield will require vast arrays of 'sleeping' rules awaiting the right trigger at the right time, and they could be deep down the levels of a vast decision tree.

In a well-confined area of application, especially in a business context, the most basic objects of interest will often be easy to recognise—parts, personnel, tasks, customer orders are examples that most experienced systems analysts will recognise as likely candidates. Building knowledge around such primitives or atoms promises to at least start off as a well-ordered affair, just like the design of file records in fact. But primitives are not so easily recognised once we move into knowledge bases concerned either with a wide variety of object types or those concerned with nebulous (literally 'cloudy') objects such as, say, news events, opinions, gut-decisions and so on. And we can make a parallel point about relationships between objects. In some areas these may well be relatively clear, consistent and stable—the permanent hardware of a railway network for example—and in others not at all so—vehicles in a city road system.

Finally, there is the nagging question of the impossibilities and incompletenesses of reality. Impossibilities tend to arise out of human behaviour itself, so knowledge bases trying to map closely human-related real-world knowledge may time and again come up against contradiction. Take for example a debt monitoring system; it is more than likely that responses from debtors contain deliberate untruths about facts or intentions of paying, and the feeding in of data from these replies into a knowledge base would need somehow to cope with this. More mundanely, errors can look like impossibilities and any experienced systems analyst must know that errors will arise however tight the controls.

Incompleteness is another can of worms. As with errors, incompleteness is an unavoidable reality in all systems of any significance. Simple logic has great difficulty in dealing with 'absent facts and rules'. If they are not present are they therefore FALSE? Sometimes the answer is very obviously yes, as we would assume is the case when we look at instantiated values—if the name is 'Smith' then surely the absence of all other names means that 'name is [not-Smith]' is FALSE. But even this is not always a reliable guide to incompleteness. A specified date for Smith's marriage does not mean that there is no other possible date for marriage since re-marriage is possible. Of course there might be clues that enable us to recognise this situation, but this implies the need for a more sophisticated knowledge base. Where an inference engine cannot find a result to tie in with a set goal, the ideal is that it should report a Dalek-like 'will not compute', or perhaps, less cryptically, 'unable to answer'. But in practice that can be a very difficult uncertainty to be sure about!

With these difficulties now in mind we can look at the various approaches to knowledge representation which have in recent years begun to gain some currency. If it means that we set off with something of a sceptical view of things then this is no bad thing at this stage of the history of information technology. The systems analyst should not pick up software tools in this area without being fully aware of the limitations due to these being relatively early days in the field of applied knowledge processing.

9.1.4 Review of approaches to knowledge representation

Approaches is a suitable term because we are looking at groups of overlapping techniques rather than at discrete ones and at methods that often are, in any case, combined in particular software tools. But the grouping is useful not only because it is the one more often than not actually used in texts on knowledge representation and allied topics, but also because it helps give shape to what can otherwise be a confusingly diverse subject with many possibilities for going off at tangents.

We shall adopt two differences from the norm however, one of sequence and the other of additional content. The additional content is that which is concerned with data modelling techniques which are well established as systems analysis development techniques and database design tools. These are not commonly placed into the knowledge representation context, at least not from a systems analysis point of view, partly because of the 'divide' we have already referred to between traditional information systems analysis and the more research-oriented knowledge engineering field.

It is usual, apart from this difference of content, to present knowledge representations—once basic logic modelling has been dealt with—by starting with the most general of approaches which is called *semantic network modelling*. We prefer to start with the **frames approach** because of its natural appeal to minds already used to information systems using files and databases as stores of knowledge. This makes a more natural progression from the data modelling start. We then return briefly to the **rule-based approach** which has already emerged from the logic-based approach of this text. We then turn to the generality of the **semantic network approach** to look at why this is the longest established in research terms, yet least well established in applied terms.

9.2 Data modelling

9.2.1 History

From the first, data processing was concerned with the creation, processing and maintenance of computer-readable files. These files were inevitably strictly structured affairs, being collections of records with rigidly defined formats with respect both to the data items or fields and to the symbol content of those data fields. The rigidity grew initially from the physically constrained reality of 80-column punched cards and carried over into the 80-character blocks of magnetic tape and disk storage. The tape and disk storage media freed the record size constraint of 80 characters so that the *physical record* of a block of data was no longer a determiner of the size of the *logical record*. It was even possible to contemplate variable-length records, where format was adaptable, at least with respect to number of characters in a field and sometimes with respect even to number of items in a record.

Where was knowledge representation in all this, if anywhere at all? It was minimally present but definitely present nevertheless. If a built system is seen to comprise a number of major files such as Customer File, Orders File, Product File, this is evidence that a knowledge representation design decision must have taken place at some point in the design. The implicit design decision behind a Customer File is something like:

> We know we have a large number of people or companies whom we can refer to collectively as Customers. Although each customer is in fact different we believe it is possible to stereotype a typical customer in respect of the pieces of information we want to keep about them. That stereotype can comprise a list of data item names and here is the list... . Furthermore the values domain for each data item can be stereotyped as comprising a certain number of characters and/or symbols of a particular type or types.

This is a fairly significant design decision (or set of decisions). We have achieved a formwork into which apparently *all relevant knowledge* about customers can be poured! This is very tidy and economically efficient because all references to this knowledge base can be programmed to know what to expect when they tap into it.

What price have we paid for this remarkable efficiency? In a nutshell, we have paid the price of not being able to 'know' our customers except in so far as they can be manifested as a set of data values which fit our record framework. (We are assuming that no separate file of exceptions is being maintained.) An early systems analyst might well have said that this is hardly any price at all, because obviously customer information is always neatly categorisable in this way. But a good salesman, even of those early times, would no doubt say wryly: "You don't know the hundredth of it, mate." Because that salesman would know many, many bits (and titbits) of information about each of his customers. Whether or not he wants to be able to put all that juicy data into a computer-based knowledge base is beside the point. The knowledge simply won't go into such a system and that's that.

Of course, record formats can be extended to include more and more detail, but large records are difficult to cope with conceptually and operationally, and practical bounds to record size are soon set by systems designers. We shall see how this problem is addressed by the frame concept in a latter section.

As data processing systems grew in size this limiting simplicity of file structure was overtaken by other events. First of all the need to cross-refer between the files became more obvious as users demanded quicker access and more meaningful response to enquiries. With greater capacity of on-line storage devices such as disk packs and large fixed disks, the files were brought into physical juxtaposition and pointers or links set up between the record types to speed up the searching processes—and databases were born! In due course this mess of ad hoc design was tidied up by the introduction of software known as database management systems (DBMS) which introduced strict rules about the families of records set up by the links and required parent records to head the chains of child records.

At the same time more and more knowledge became literally programmed into the systems in order to handle the increasingly sophisticated processing demand of users and of the infrastructure of the systems themselves. Vast amounts of such knowledge is by now hard-set into programs which, although they may work well enough in their own way, can only be gazed at through very thick plate glass—it cannot easily or cheaply be touched and picked up for use in other ways.

The later, but eventually parallel, development of relational database systems promised some relief on the question of user access to data. The hard-wired links of hierarchical databases were placed by the apparently unlimited relationship-making facility between data relations. A user query triggered logical connections which didn't rely upon there being chains of or families of records already in existence. Apart from the fact that some of the promise is purely illusory, the reality of how much knowledge can be stored by such relational data is far short of satisfactory when looked at from the knowledge engineer's perspective.

This potted history has been potted because it is to a certain extent a diversion from the theme of knowledge representation. It has been very largely to do with the attempt to improve the handling of data forms as simple-formatted records, not with the changing of those forms to improve their knowledge-carrying capacity. However we have so far left out an important recent episode of the story which tells how all this concern with the design and structure of databases led to the development of what is known under the general heading of **data modelling**. Data modelling techniques have nurtured some sophistications of method that are, at last, worthy of being considered as candidates for knowledge representation. We shall now turn to data modelling therefore as a way of completing this potted history.

9.2.2 Entities and attributes

The term **entity** as used in data modelling (it is a word that pops up very often in computer-based information studies, and with a variety of, sometimes subtly, different meanings) shows very clearly its ancestry in file and database terminology. It is really an abbreviation for *entity type* because it actually refers to a collection or class of objects, just as do a record type and file of such records. In fact when looking at entity types within a database the synonymity of meaning with record types is almost totally complete. Not quite though, because 'records' will be used more often than not to emphasise reference to the physical reality of actual data.

The emergence of the word entity seems to have been to do with trying to shake off this more physical record image for the purposes of analysing real-world data without commitment to the reality of records. Thus the systems analyst at early stages of investigation can refer to the dawning of awareness of classes of data as **candidate entity** types. 'Personnel' or 'product' may be immediately obvious candidates, but others may take a good deal of deep understanding about the system to emerge. More abstract types such as events (eg. message receipts) or information-based data (eg. re-order requirements) are typical of this type. This stage of analysis is sometimes referred to as *conceptualising* and the entities at this stage as **conceptual entities**.

This type of conceptualising is done very much with the end-product of database record types in mind, and this anticipatory frame of mind may be said to distort the dispassionate and balanced detection of 'true' entities. Be that as it may, it is only a reflection of the pressure on systems analysts to come up with implementable designs in commercially viable periods of time.

The **attributes** of entities (again *attribute type* is perhaps more accurate) are the potential precursors of the fields or item types in records, and systems analysts will again be conscious of this at fairly early stages of conceptualising. But at conceptual stages the question of what should and what shouldn't be attributes is more acute since the concept of the entity (what actually is a 'personnel' exactly?) is directly manipulated by the inclusion and exclusion of candidate attributes. Bunches of attributes may be seen to be candidate entities as analysis proceeds and the question then arises as to what precisely *is* an attribute.

An attribute is not in fact a precise concept. One may attempt definitions such as: a simple data class which has no meaning independently of the entity to which it is attached. This captures the sense of its special association with the parent entity, but can soon be shown in general not to be uniquely attachable to a given entity. Uniqueness may be forced upon the association when it comes to implementable designs—clues will be found in the allotted synonymic names such as personnel_name and manager_name—but this belies the conceptual reality of entities often sharing attributes. Indeed in relational database records, shared attributes are essential to getting any results from the system at all.

But the identification of entity/attributes classes is not independent of the perceived potential relationships between entity types anyway, and we must therefore widen our picture to include the **entity-relationship** (ER) view of data modelling.

In the most general sense a relationship refers to a connection of some kind between two things. Thus in the entity/attributes model we have already implied two kinds of relationship.

First, by referring to types or classes there is an implied relationship of a 'X is an instance of Y' kind (we shall use IS_A) between an individual member of the set of things and the class description. Thus a data value is in an IS_A relationship with an attribute name, and an entity occurrence (a complete data record) is in an IS_A relationship with the entity name (see Fig 9.1). In actual database systems these names will be part of a descriptive schema and the IS_A relationships are embedded in the very working of the DBMS software. Neither in the design specification nor in the implementation do the systems analysts or programmers have to create these types of relationship other than by declaring the names in the schema and feeding actual data and records to the DBMS when creating the actual database.

The second kind of implied relationship is that between attribute and the entity to which it belongs. This is a HAS_A kind of relationship (reading in the direction from entity to attribute). It exists both at the names level and at the actual data values level (see Fig 9.2). Again the schema and the DBMS software supporting it looks after these relationships as being part of the intrinsic record-like structure of things. This implicitness is not adequate when we take a more general view of knowledge modelling

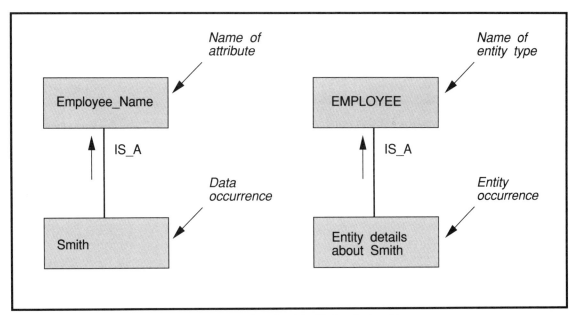

Fig 9.1 Data occurrences involved in IS_A relationships

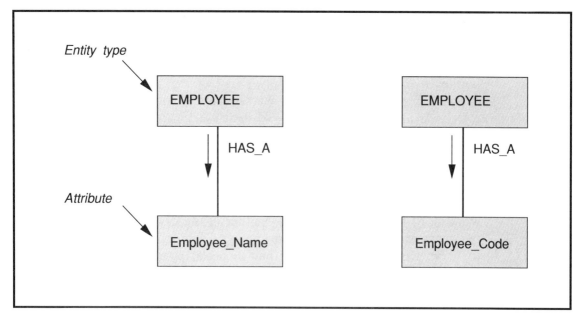

Fig 9.2 Attributes involved in HAS_A relationships

requirements, as we shall see later. However, it is what is expressly and universely provided in database systems and what therefore most systems analysts are used to. Furthermore they will expect these basic structural relationships more often than not to be manifest as physical juxtapositions of the data, both as declarative descriptive data within the schema and as value data in the database itself. Sets of entity records will be arranged like files, though there may be pointers or links to support the physical continuity in certain circumstances, and records will usually comprise data arranged contiguously in the storage medium.

The kind of relationship that almost all systems analysts will have heard of and actually be familiar with is the kind that occurs between entity and entity. What these *entity-to-entity relationships* should signify is rarely conceptually constrained. It is more likely that the decision to incorporate a relationship will be based on pragmatic considerations about how the data is likely to be used in practice. In an interpretation of such modelling comparing it with the file handling traditions which preceded its development, one would see these relationships as embodying what would have been cross-references between files. That is to say that previously this would have been achieved through a program or sub-routine searching a second file on the basis of a criterion or criteria presented from a first file. For example,

Given that Smith's Personnel record shows that he runs a company vehicle, find all data concerning the maintenance of all vehicles which

Smith has run by searching the Vehicle Maintenance file, using Smith's identity as search criterion.

If a PERSONNEL-to-VEHICLE_MAINTENANCE relationship is established, the DBMS will initiate the search as a standard part of its service. The relationship in this example might be named something like HAS_RUN_VEHICLE, reading it from the PERSONNEL Entity to the VEHICLE_MAINTENANCE Entity direction. The special name for it suggests that the relationship itself is a piece of information. Is there any data that could be attached as attributes of the relationship? The most important and familiar relationship data is that concerning its degree.

The **degree** of a relationship expresses the type of enumeration that can be expected when actual instances of the relationship are triggered (as for Smith in our example). The example we have taken would suggest, for example, that for each PERSONNEL occurrence there could be zero, one or an indefinite number of related VEHICLE_MAINTENANCE occurrences. A relationship is said to have a degree $m:n$ where the m and n symbols may be just that or may be replaced with 1 where singularity of connection is guaranteed. In diagrammatic terms our example relationship might be expressed as shown in Fig 9.3.

The n indicates the indefinite-number end of the relationship degree (the individual may have run a number of vehicles) and the 1 indicates that any particular vehicle has only ever been run by one individual. The broken line indicates that the connection is optional—in this case meaning that there are

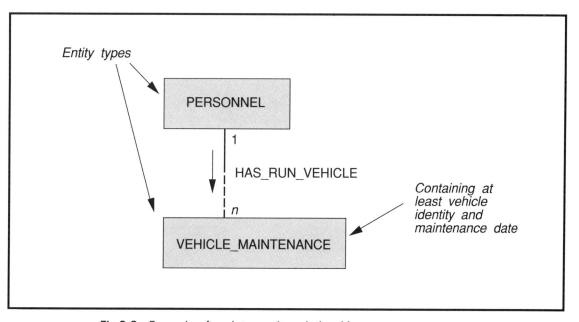

Fig 9.3 Example of an inter-entity relationship

263

instances where no vehicle has been run by an individual and that therefore the relationship does not need to exist for that individual.

Let us imagine that further analysis provides new evidence. First, that vehicles may after all be run by more than one individual, and second that since each vehicle is maintained on several occasions there is need for a further entity, MAINTENANCE_EVENT. The diagrammatical model—called an *ER Diagram* or sometimes (rather confusingly) Logical Data Structure Diagram—might now look like Fig 9.4. (To keep things simple, optionality has been dropped.)

The HAS_RUN relationship is now shown as *m:n*, usually read as many-to-many. This may reflect clearly enough the kind of connectivities that may be found between the two entities, but often in data-modelling good practice, it is recommended that these rather complex connections be simplified by interposing a linking entity as shown in Fig 9.5. Also we have added a new relationship between PERSONNEL and MAINTENANCE_EVENT to enable connection to be made between the maintenance event and the individual who was running the car immediately prior to this maintenance. The event is thus associated with one individual, though an individual will be associated with zero, one or many maintenances.

What has emerged is the beginning of a network ER structure, with entities as nodes and relationships as links. Database systems based on the ER approach are referred to collectively as **network databases** in fact. The network structure is interesting from a knowledge modelling point of view, as we shall see when we turn to semantic network models in due course.

There is a great deal more to ER modelling than we wish to go into here. There have been varieties of notations and rules about how the models should be specified, but most CASE systems will include a form of ER

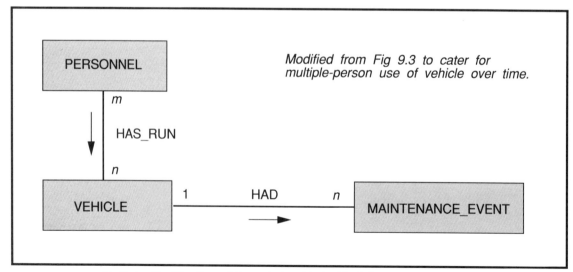

Fig 9.4 Modification of the relationship in Fig 9.3

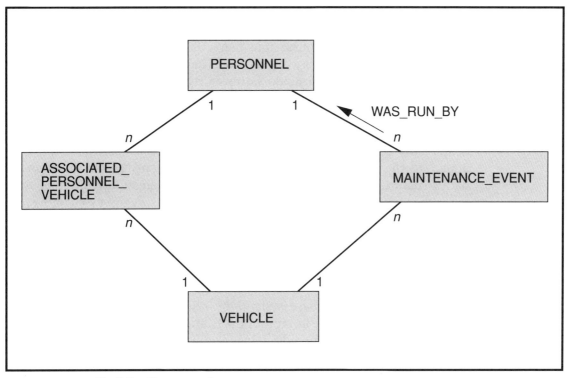

Fig 9.5 The start of an Entity-Relationship network structure

modelling as one of the incorporated tools. This suggests that the technique has earned wide respect as a genuine development aid to systems analysis, and this is indeed the case. We can therefore conclude that it does succeed in providing its user with some effective means of representing knowledge about data structures.

As we have been suggesting, this is not really fresh news to experienced systems analysts, but we have gone through the steps of a mini-review by ER modelling for this very reason. It seems quite important to take the opportunity to give systems analysts the confidence that they do already know something about knowledge representation—and we can now make use of this start to see why more advanced knowledge representation techniques are needed when it comes to modelling knowledge of a less specific kind than that related to data structures.

9.2.4 Relations again

Before turning to these other approaches, we ought to return for a moment to the question of relational forms of data modelling and handling which we have already looked at in Chapter 5.

An entity as we have presented it here is a type of relation. It is an n-tuple relation where n is the number of attributes. The attribute names are the data-set names. We did in fact point to the correspondences between relational forms and data processing files in Chapter 5, and similar correspondences will of course apply here. There is a great deal of common ground at a basic level between relations, entities, tables and records.

Relational database forms do however rely upon relations having data domains in common in order to make the equivalent of relationship connections using the JOIN operators. This is one of the crucial issues at the heart of the 'network versus relational database' debate. Data modelling at the early conceptual stages can be used irrespective of which approach applies to the end-product database, but the relationships displayed on the data structure diagram are made tangible in fundamentally different ways in the two approaches of network and relational. As we discussed above, the relationships are explicitly specified in the schema of network systems, and this enables control and other data to be attached to these data objects. Degree, optionality and storage modes are among the features that can be given constraint values to guide the DBMS in achieving accuracy and efficiency.

For a relational database system the relationships have to be set up implicitly by nominating an appropriate domain to play the commonality role. The suggested common domains for our model in Fig 9.5 are indicated in Fig 9.6. Some of these would have been present in any case—the ability to identify the individual would have been essential in the Personnel entity, as would the need for the vehicle identity in the Vehicle entity. But it would not have been obvious that the personnel number should be in the Vehicle_Maintenance entity. We also find now that it is not necessary to have Personnel Numbers in the Vehicle entity. This aspect of modelling can quickly become complex and confusing and in practice the analytical technique known as *normalisation* of data is required so as to arrive at data domain inclusions that are rationally related to the required data model. (We shall not deal with normalisation in this text.)

COMMON DOMAINS	
Attribute	*In Entities*
Personnel_ID	PERSONNEL, ASSOCIATED_PERSONNEL_VEHICLE, MAINTENANCE_EVENT
Vehicle_ID	VEHICLE, ASSOCIATED_PERSONNEL_VEHICLE, MAINTENANCE_EVENT

Fig 9.6 Selected common domains for the E-R model in Fig 9.5

An interesting problem associated with relational forms is that, if data domains are not defined tightly enough, entities might have common data domains which do not support sensible relationships. Thus a data domain of 'year as a six-digit number' might appear as 'year of birth' in one entity and as 'year of manufacture' in another. A JOIN between these two entities is a practical possibility but unlikely to have much usefulness or even meaning as a relationship. This implies the need to define at least two different domains for dates, which will almost certainly need to have different validation checks applied to them in any case.

The implicitness of relationships in relational forms as opposed to their explicitness in network forms suggests that they are a disadvantage in working environments, and it is certainly true that systems analysts steeped in the network culture remained sceptical about relational database forms for many years, especially with respect to questions of processing efficiency. The debate is an important one to the knowledge representation area because, while network systems may indeed have practical advantages for the commercial database designer, the relational forms are more soundly tied to formal logic. And since formal logic has wider currency across all approaches to knowledge representation this must put it in a stronger position when considerations of interfacing between knowledge processing systems and database systems are taken into account. However, the debate does not end there, since the emergence of hybrid systems will almost certainly show that the advantages from both can be brought together.

9.3 The frames approach

9.3.1 From entity to frame

We have already discussed how an entity is a collection or class of things and how conceptual modelling helps the systems analyst towards selecting these entity types. We have also seen how the next step in this identification is to attempt a specification of this entity type by means of a list of attributes which have to be simple data-item formats. The formats have associated with them certain constraint rules which are designed to ensure that only valid data values get into actual entity records. Entity types may then be linked to other entity types by means of named relationships, which in turn have constraint rules associated with them. And all this information about the entity, to be found in the schema of database systems, forms a kind of framework for each set of actual data to be shaped by.

However, it is not an actual framework but possibly a rather scattered collection of schema statements. It is certainly not addressable or retrievable as a single data object. The first step towards the frame approach is then to provide a knowledge-based system (KBS) which does treat all this information about the entity type as a cohesive frame of specifications for, and rules about, the data items. (Frames are also sometimes called schemata.) This should be a straightforward enough concept for database trained systems analysts to handle, though it is only a first step towards the frame approach.

9.3.2 Slot-and-filler structure

If we take the frame concept as being merely this gathered-in collection of schema statements, we have a somewhat heterogeneous set of items to deal with. We then have to face the problem of what shape to give this framework. Do we have a section for data item descriptions and validation rules and another for relationship descriptions and rules? In theory the answer is that we do not. Instead we agree to have fairly unconstrained slots into which these things can be placed. Rather than being a definite section of storage, the slot is just another bag waiting to be filled with whatever seems appropriate. Hence the phrase *slot-and-filler* which is often used to convey this arrangement within the frame approach.

The slot-and-filler facility enables us to contemplate including other kinds of information about an entity type. The most important of these is the incorporation of the idea of *inherited* data.

9.3.3 Inheritance between frames

A very common feature of human-organised information is the putting of things into classes, sub-classes, sub-sub-classes and so on. In the sciences, especially biology and zoology, in library book coding, in manufacturers' catalogues, in management reporting structures and in structured analysis and design techniques, and indeed in many other contexts, the use of a hierarchical scheme of ordering is the back-bone of the structuring of the data. The first types of database system used hierarchies as the main scheme for allocating data to storage space, and many still do.

There are psychological, even philosophical, explanations for this instinct to organise into levels and it is not surprising that knowledge processing theories should also look to this method for organising knowledge. It is a feature which should be a staple availability even if not used in any particular case. And a powerful justification for saying this is that in any context which aims to store information for future retrieval there is an implicit need to access other information up and down the tree structure from any data item in the structure.

When the hierarchy is a strongly connected one such as in classification systems, so that the relationship between an object at one level is such that it is A_KIND_OF thing at the next level up, it is more than just a case of access. It becomes a case of legitimate need to inherit data from the higher level to the lower in such a way that the lower level incorporates that data as part of its set of attributes.

Take, for instance, a classification of road transport vehicles such as shown in Fig 9.7. Information held at the vehicles level (eg. power = automotive) can also belong to all items in the tree below it. But it only needs to be literally inherited one level at a time, since the inheritance chain can bring data down all the way from the top. The power = automotive data can be inherited by every level if necessary. There can of course be maximum levels above which data will not be found. Thus Cavalier L can inherit information from Cavalier (eg. about body style), Cavalier can inherit information from Vauxhall (e.g. HQ address), and so on, but in each case the chain stops there.

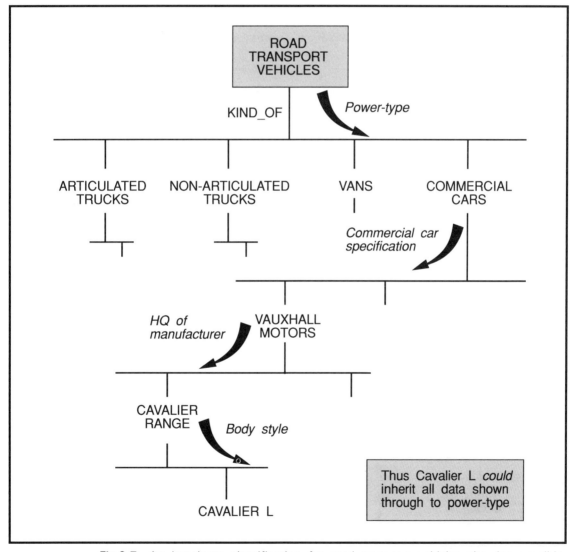

Fig 9.7 An imaginary classification for road transport vehicles showing possible inheritable data

Another strongly connected hierarchy is the type associated with component structure in mass-produced goods, where the inter-level relationship is PART_OF (or HAS_PART reading down the tree). We can again take vehicles as an example (Fig 9.8). Note that each level here has correspondence with separate physical objects—it is not just a question of the convenience of naming groups of things. The inheritance need will now be slightly weaker however since more information at a given level will be concerned with that level and that level only (eg. weight). Other information (eg. location within vehicle) may be deemed inheritable, and this draws our attention for the first time to the need to be able to distinguish between these two types of information—inheritable and non-inheritable.

Before we address that need we must take note of another aspect of the inheritance of information. We know it colloquially as 'the exception to the rule'. Smith, shall we say, is running a vehicle which is an actual instance

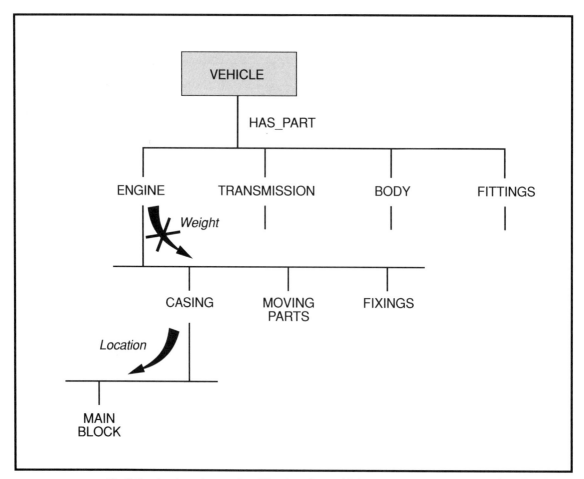

Fig 9.8 An imaginary classification for vehicle component structure showing both inheritable (*Location*) and non-inheritable (*Weight*) data

of the Cavalier L, and it has specific attributes such as colour and seat fabric which rightly belong at the level of actual vehicle. But it would be wrong for it to inherit the fact that it has stereo radio cassette player (which is standard) because it so happens that Smith had that equipment removed—he required the space for special storage shelves to go with his use of the vehicle for sales support work. Our attention is drawn here to the need to distinguish between data inherited by default and real-world data which cancels this default value.

Both these requirements point to the need for a facility in the frame-approach knowledge system to mark attributes as being either cancellable or non-cancellable as inheritances from the parent level by default. Non-inheritance can be dealt with simply by absence of any inheritance marker. Figure 9.9 shows part of a frame for Smith's vehicle showing which attributes are which in this respect.

Note that it would appear that all frames at this level must have the same categorisation of cancellable/non-cancellable default values (radio/cassettes must be cancellable for all actual vehicles). This can create a knotty judgemental problem if, for example, only one vehicle requires the cancellation in practice and the two thousand other vehicles should definitely have the standard. To the knowledge theorist it poses an even

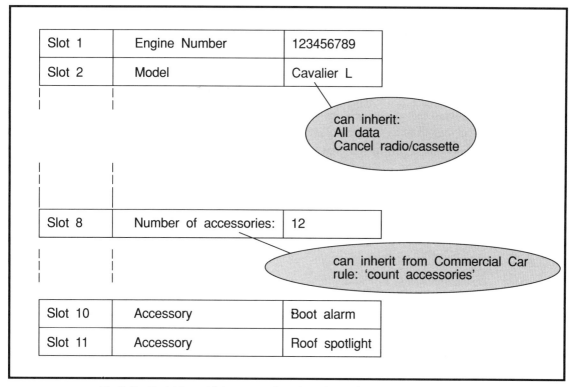

Fig 9.9 Smith's vehicle: part of an imaginary frame occurrence

more serious question, namely how could an inference engine make the common-sense deduction that (usually) all Cavalier L models have radio/cassette as standard? It could report a statistical result of course, but how could that result be used in further deductive reasoning? We shall touch upon such issues in the final chapter.

9.3.4 Rule-holding slots

We have already touched upon the possibility of keeping validation rules alongside the data. Any other type of rule is a candidate for inclusion so long as it can generate data which belongs to the frame in which the slot resides. For instance 'total number of accessories' could be stored as a search-and-computation rule rather than as an actual value—at the model (Cavalier L) level because this may change from year to year and automatic updating of the knowledge is clearly desirable, and at the actual vehicle (Smith's) level because the cancellable defaults facility will mean uncertainty about the total until the frame is a concrete reality (see Fig 9.9). Notice that here the rule has been inherited by this level from the parent.

Although in theory any rule can be incorporated, care has to be taken to ensure that it does apply to all siblings (frames at the same level sharing the same parent) unless rules too can be made cancellable. Rules should always of course leave everything except the slot-filling result unaltered.

9.3.5 Frames and object-oriented programming

Although object-oriented programming (OOP) has wider application and relevance, there is a close connection between OOP and the frames approach. A frame necessarily constitutes not just a static schema-like structure but also by implication it is a small library of software! And yet we require a frame to be handleable as a single coherent data object too. Treating a chunk of software as an independent data object for processing use by other software is what OOP is chiefly about.

Some would see it as inevitable in the presently burgeoning era of CASE, 4GLs and re-usable software that OOP will become the accepted standard style for software design and development. How this standard can actually be defined had not yet fully emerged by the end of the eighties, but it seems not unreasonable to predict that when it does there could be a significant boost to frame-approach KBS.

9.3.6 An overview of KEE

We mention KEE (Knowledge Engineering Environment) here because it is the KBS probably most closely associated with the frame-based approach. Developed in the early eighties, by a firm now known an IntelliCorp, for building hierarchically structured knowledge systems, it has been tried more recently on a wider basis, especially in the area of planning. We shall attempt neither a detailed description nor an evaluation, but use it merely to indicate how the ideas we have been discussing have been made available in an actual software package.

It provides a slot-and-filler frame basis more or less as we have described these, except that the slot should be read as an even more sophisticated storage availability than we have perhaps suggested. Its especial provision for hierarchical relationships and the inheritance principle suggest that most is to be got from it by building the knowledge top-down. The higher-level frames, once defined, enable labour-saving definitions at the lower levels because inheritance is a principal feature. However KEE also incorporates a rule-based inference engine that works with all rule-carrying slots using the rules of logic we discussed in earlier chapters, though it appears to use a more general form of resolution than Horn clause formats. Another important feature is the provision of active variables which are guaranteed to be processed dynamically as their values change—an essential requirement when modelling systems where the basic data changes continually such as in real-time systems.

Where the inference engine finds it is unable to process a query due to lack of sufficient data it may choose to prompt the user for that data. It will also respond to the user's Why? by explaining the line of reasoning leading to the blockage.

Of major importance with any KBS is the nature and quality of the user interface. This is represented in KEE by means of some highly sophisticated windowing software (and therefore requiring the more powerful types of equipment). Apart from the fairly standard input–output window features associated with any window-based personal computing package, there are special features such as text-and-line displays of the hierarchical structure, text-and-line displays of the network of rule interrelationships, and dynamic graphical displays of the activity variables (the pictures change as the values change) in pictographic forms such as dial-face instrument display, scale-and-pointer display, liquid-in-tube pressure display, as well as the usual range of statistical graph types.

KEE is no 'toy' system and its sophistication means that a user needs not only sound training in the manipulation of the interface and so on, but also a less-than-superficial understanding of what the package can do and what is going on when it does it!

9.4 Rule-based approaches

9.4.1 Introduction

Having looked at inference rule forms (if … then …) from almost every possible angle it might seem there is little left to say at this stage that won't simply tend to confuse. But although we have looked hard at the meaning of material and logical implication, we have not established what are the actual features of a rule-based approach.

Put in the simplest terms a rule-based system invites its user to fill in the missing parts in sentences of the general form:

IF ... AND ... AND ...   THEN ...

It assumes the user knows how to obtain and formulate the knowledge so as to insert into these gaps information that the interpreting part of the inference engine can make sense of. There is a variety of appearance in the interfaces of the different systems which we shall not attempt to review here, but the potential user need only be aware of this simple structure initially.

Indeed this simplicity explains why expert system shells have proliferated in recent years. The demands on the user to supply the material for the above gaps seem temptingly easy to meet. Used within their limitations these tools do provide means of programming the knowledge contained in such applications as advice and diagnostic systems, so long as the knowledge domain is clearly bounded and has associated with it an 'expert' capable of yielding up the rules (with help, perhaps) from current thinking.

But there are limitations. Actual expert systems have so far tended to be built from scratch, so that the conditions that are to be searched out in order to test whether a rule should fire or not are based on facts put into the system as part of the whole knowledge acquisition exercise. The medical diagnostic expert system is built with a fact base expressly designed for it; the timetable advice system is likewise loaded with timetable data known to be interpretable by the system. But more general expert systems which might for example be scanning the performance of a world market will need to be able to access databases from a wide variety of types and quality. Under such conditions a simple rule-based expert system will soon find its reasoning at best flawed, but more probably floored!

We shall therefore consider some of the implications of such issues and then look briefly at a sample expert system builder which is accessible to the modestly resourced user.

The reader may have come across from time to time the expression 'production system' or 'production rule system' in connection with this area of work. The term was originally coined to refer to early research work in such specific fields as the logical formalising of rules of grammar—thus the grammatical rules would 'produce' actual sentences when given words and phrases. The terminology does not seem particularly useful now, but it has to be recognised that it has stayed in common usage in some circles, especially when it is being implied that a rule-based system produces usable results. The important point is that the reader should not take it to mean anything more than it does.

9.4.2 Closed domain and closed worlds

When we look at

IF [condition] THEN [action]

we can quite reasonably interpret [condition] as meaning that whatever it is it has to be TRUE for the rule to fire and the [action] put into effect. If it is a simple Boolean variable or expression there is no problem. But if it contains criteria that involve matching it with stored data in a database things are not so straightforward.

We need to consider a problem we have touched upon before—the problem of closed data domains and closed worlds. In a simple Venn diagram view of all the facts in a database, the whole database is contained within one circle which says effectively 'everything within these bounds is TRUE' (see Fig 9.10). But what is the significance here of the rectangular boundary? If it represents all possible facts within the context of the use of the database, a serious problem arises in respect of the truth value of the facts not within the database enclosure. A common-sense interpretation would be that we do not know yet about these not-gathered-in facts. The Boolean interpretation is however that they are FALSE, because in a closed world that which is NOT TRUE must be FALSE.

Thus if a rule-condition requires a reference to the database to the effect that

IF salary of Smith is greater than 20000 THEN...

the rule will not fire if Smith's record is not yet on the database. This might be a nuisance if this is due simply to some delay in input of data but no

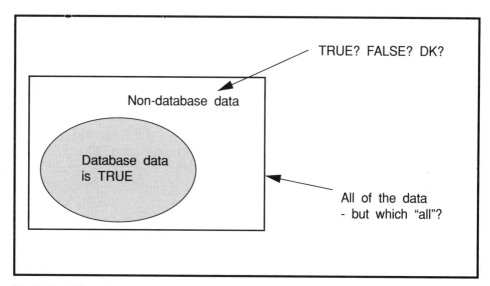

Fig 9.10 What about the data outside the database?

logical damage has been done. If however the rule contains

IF salary of Smith is not greater than 20000 THEN ...

the rule may fire if the interpretation is equivalent to

IF NOT salary of Smith is greater than 20000 THEN ...

because the condition that follows NOT is FALSE and therefore the condition including the NOT operator is TRUE.

The answer to the problem here would be to incorporate some kind of integrity constraint on the personnel domain which would recognise absence of related data (the salary here) as just that, and not interpret it as FALSE. Indeed actual programs written to enquire of such a database would almost certainly take that line. However, some absences of data do mean that the data should be taken as FALSE. For example,

IF NOT Smith is an employee THEN ...

Here, Smith's absence from the employee domain does mean that 'Smith is an employee' is FALSE.

The problem can occur in any database system if there is undiscovered confusion over domain integrity constraint. For example, a ferry berth-booking system refused to allow a tour party to return with an additional member because of its validation check failing, even though the family had genuinely picked up a relative for a visit home.

The same problem can arise in connection with not-yet-defined relationships in network databases. It behoves the database user to know that relationships exist and are active in a particular database before setting up queries. Again, integrity constraints can act as sentinels to prevent non-viable queries from being put to the system. Since here the question of where the rectangular frame closure should be set is much more open, the problem is termed the Closed World Problem (CWP) and is one that is still receiving the attention of researchers and software developers.

A further domain-related problem arises with the merging or parallel-usage of two or more databases. Unless there is a higher-level control of domain constraint definitions, synonyms can be introduced which, while being bad enough for ordinary database use, can be disastrous for rule-based systems usage. Imagine 'red' being used for vehicle paint colour in one database and political colour in a second. The 'jokes' about being able to conclude that all our fire engines have leftist tendencies and so on are legion.

While software which has been written expressly to work with databases can be and usually is provided with appropriate integrity constraints, the interfacing of rule-based systems with databases is much more problematical and the systems analyst should show due caution about would-be enthusiastic knowledge system vendors who suggest that their software has solved this interface problem.

9.4.3 The conflict-resolution factor

As explained in the previous chapter (Section 8.3.1) there can be enormous numbers of options open to an inference engine in terms of choice of computational sequence. An allied issue is the rather formidably named *conflict resolution strategy*. This refers to the actual way or ways in which an inference engine selects which rule to fire when there are many current candidates. Again we may seem here to be at the boundary of "Do I really need to know this?" as far as the systems analyst is concerned. But there is an aspect which has a direct bearing on the actual use of any rule-based software system.

Because the options can again be large in number the engine may simply take the next in line, based on the sequence as presented by the programmer of the rules. This means for example that in our Prolog example the solutions pop up in a sequence determined by the sequence of the Prolog Clauses given by the user. If there are too many solutions to justify full computation the system may stop without having reached the most interesting solution. Of course the user can change the program, either in terms of sequence or in some way that narrows the goal and consequently the search. But there is a certain arbitrariness about this which indicates that in practice the action of a rule-based system is not independent of the way the program is written. This is no surprise to programmers experienced in procedural-type languages, but at least with such languages the control is more direct and predictable.

The inference engine may be capable of using other selection criteria such as giving priority to most recently added rules, priority to rules with most (or least) consequence, or priority based on context-sensitive information including direct updates from the user. This last type takes us into the area of rules about rules, or meta rules, and implies that the system builder(s) know a good deal about the design principles of the inference engine. This kind of tuning facility will certainly need to be available for developing and running large and sophisticated systems, but whether systems analysts will need be involved to this depth seems unlikely unless specifically allocated to this type of work.

9.4.4 Crystal—an easy introduction to rule-based shells

Crystal (see Section 6.4.4), developed and marketed by Intelligent Environments, is a software package for capturing, storing and processing knowledge as sets of rules. The interface is a friendly one and even a naive user can make sense of its purpose and function in a relatively short time. The building of fully fledged Expert Systems is within the capabilities of the software, though naturally much depends upon the ability and understanding of the user when it comes to developing truly useful systems.

The user adds rules to a knowledge base via a window and a menu-presented command set as with many other personal computer packages. At the heart of the business though is the simplicity with which the rules can be set up in the window with a minimum of structural effort on the user's part. (See Fig 6.11.)

The first line is automatically offered with a prefixed IF and the user then inserts conditions which can be connected with AND, OR and NOT by one or two keystrokes. Appropriate actions are then linked to this condition structure. The condition formation is supported by a system of variables which the user defines and manages in a relatively natural way.

Starting with a package like this is really as painless as it sounds. The pain will actually have begun when the user had to formulate the system to be modelled in the first place. Indeed knowing where to start with a candidate system (or even finding and recognising the candidate system) will prove to be the trickiest problem, no doubt. Here, the software is not a lot of help—though one is able to view what one is in the process of building.

Whatever the limitations of such a tool it nevertheless provides a useful way in to the basic idea of building a logic-based expert system and as such may be regarded as a valuable introductory learning and training tool.

9.5 Semantic networks

9.5.1 Introduction

We have already considered the topological properties of networks (or nets) as comprising a complex of interconnected arcs (or links) and nodes, and we have used such structures to model certain conceptual ideas, data relationship modelling in particular.

But what of the prefix 'semantic'? Most reasonably well-read people will have at least a vague notion that semantics is something to do with meaning in language. Every word in a natural language has a cluster of meanings associated with it. In computer programming languages it is just about possible to claim that there are key reserved words within given languages that have only one meaning—the meaning they were defined to have when the language was designed. But that claim can look decidedly shaky in an actual human context—try asking three very experienced Cobol programmers to define the meaning of the verb MOVE in that language and you may discover how even an officially rigidly defined word can at least sound very different in meaning when filtered through individual human minds.

But in any case semantics doesn't stop at individual word level. Words have meanings related to the context of the phrases and sentences within which they are placed, and furthermore sentences have meanings which are usually much more than the 'sum of the meanings of individual parts'.

Consider the following:

The word Break
- Operators are allowed a break at least once per hour.
- A break in the production line must be avoided.
- Given a good Summer the firm can break even.
- The Sales Manager said "Give me a break!"
- All operators enjoy a break when it happens.

We can all read and recognise the word as presented, though which of its many meanings—some of which are exemplified by the accompanying sentences—we think of first is unpredictable. While the first sentence is reasonably likely to be interpreted correctly, the next two involve some knowledge of the subject area and the last two are each open to more than one interpretation as whole sentences. The last line is dangerously open to two quite contrasting meanings if one takes the first two usage examples as guides!

This is not a 'clever' example. Choose a commonly used English word at random and anyone will find that such examples pour forth without any great mental effort. Whatever 'semantic' means it certainly can uncover a wealth of complexity and confusion. What then is it that semantic networks are supposed to be models of? Models of natural language itself? At the most ambitious level, yes, but what we have already discussed should be fair warning of the problems lying in wait to thwart that ambition.

And with just two types of element at our disposal, node and link, how can we contemplate dividing the natural language world into two classes and two classes only? Again the attempted answers to this question are fraught with tremendous difficulties. Yet with data modelling, for example, we do try to achieve just that—the data world comprises just two types, entities and relationships, though it would be a brave person who would try to claim that data models can be read just like natural language! But since our perspective is bound to be heavily influenced by network modelling with which we are already familiar, we shall start from a point not too far removed from the world of data modelling, while at the same time trying to inject the flavours of more natural language models.

9.5.2 A basic approach to semantic net modelling

Without too much philosophising we shall take a big leap into the middle of the arena and accept as a possibility that the language world can be divided into just two classes, roughly characterised by nouns and by verbs. We suggest that noun-like things are objects which generally have a substantive continuity associated with them, while verb-like things suggest what can happen to the noun-like things in relation to one another. We must not be too 'physical' in this interpretation. There will also be *abstract* noun-like things which have abstract verb-like things that can happen to them!

We have already met the relationship types IS_A, KIND_OF and PART_OF. These are verb-like things (KIND_OF is really IS_A_KIND_OF and PART_OF is really IS_A_PART_OF) and we shall now extend this list of types rather arbitrarily as follows:

A IS_A **B**	**A** is an instantiation of a class **B**
A PART_OF **B**	**A** is a component of **B**
A KIND_OF **B**	**A** is a sub-class of **B**
A IMPLIES **B**	**A** materially implies **B**
A CAUSES **B**	event **B** must follow when event **A** occurs
A EXPLAINS **B**	**A** enhances knowledge about **B**

This last type is the most general and is perhaps more honestly described as the 'all other types' category. We shall discuss this classification of relationship types critically shortly, but first let us take an example of trying to model a small part of the knowledge about an organisation. Whereas in data modelling it is usual to confine analysis to specific areas at a time in order to generate relatively consistent and cohesive models, semantic modelling should be more serendipitous and follow natural lines of thought and explanation as they might commonly occur in rational (here, business) conversation. Fig 9.11 shows a possible first attempt.

We have of course steered the exercise towards giving examples of all of our relationship types. The relationships have names which suit their function in modelling the narrative equivalent of the knowledge—the type names are given in brackets. Sometimes this involves changing the reading sequence, for example when HAS is used as a PART_OF relationship. It is therefore helpful to use small arrows to indicate reading direction.

What advantage does the recognition of relationship types give us, if any? It might be claimed that it gives some pattern to what otherwise is a confusing swathe of diagrammatic interconnectivity. Where we spot IS_A, PART_OF and KIND_OF types for instance, we know that we have a hierarchical relationship which is capable of breeding many instances to go with the parent object, and the subsequent distinction between these three types helps us to think about what data are candidates for crossing the relationship in the form of inheritance connections. The IS_A relationships in particular are very likely to involve high volumes of instances (all individual managers in our example) which are unlikely to be concerned with very abstract things.

But this hardly provides answers to all the problems with a model of this kind. It is worthwhile testing the readability of the model with such interpretations as

Smith is Senior Manager, which is a grade of employee which normally runs a Company vehicle. Smith does in fact run one—its ID is 1234. As an employee, Smith receives a salary which has two components, a basic component and an overtime component....

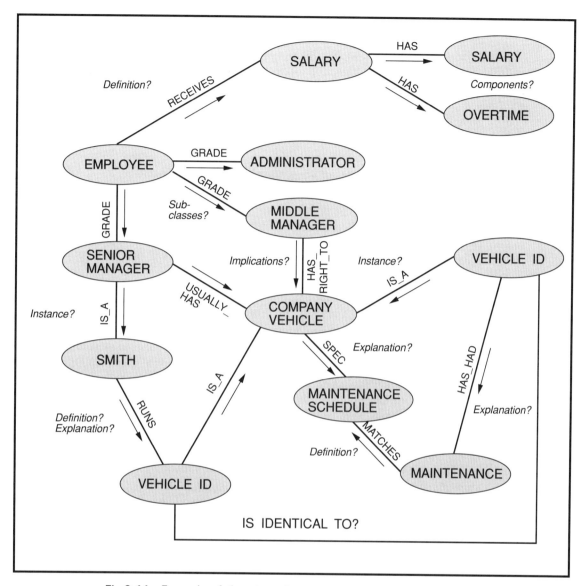

Fig 9.11 Example of the start of some semantic modelling

But what if Senior Managers are not in fact eligible for overtime payment? Or, more complexly, Smith is the only Senior Manager who does (or who doesn't, would be equally problematical). See how this kind of problem has been tackled in connection with the running of Company vehicles. The relationship is called NORMALLY RUNS to imply that the relationship RUNS between Smith and his actual vehicle need not appear against every instantiation of Senior Manager. Put the other way around though, shouldn't RUNS only be allowed to appear if NORMALLY RUNS or ALWAYS RUNS or SOMETIMES RUNS or whatever appears elsewhere between Senior Manager and Company Vehicle? Where does the model specify such constraints? Even though we are only looking at a very small portion of the total possible model it is not clear how such partially hidden logic can be modelled by the network.

We could allow that each object-like thing, enclosed in ovals here, has supporting descriptors which could serve this purpose as well as enhancing our knowledge about the object. But notice how like frames with slot-and-filler support this is beginning to sound. We resist taking this return step towards frames, however, because that would take us too soon away from the spirit of generalised modelling which semantic networking embodies. We shall instead move on to *scripts* to make this point.

9.5.3 Scripts

It may be helpful to note that an alternative name for script used by some authors is **schema**, because the systems analyst who has had to draw up schema for databases will immediately think of it as a structure-describing device. This is what a script is but in a much less formalised way than the software-supported schema of database systems.

Cognitive scientists have suggested that perhaps one way in which humans handle knowledge is by clustering it around recognisable wholes such as 'making a journey', 'eating a meal', 'using the personal computer', 'visiting a library', 'consulting with a client' and so on. It is certain that even if it is true that we do do this there are structural backdrops upon which such pictures are hung and uses of analogy, contrast and comparison to help catalogue this picture library. But forget these for a moment and concentrate on the supposed scripts themselves.

Attempts have been made to provide some kind of framework for scripts such as: themes, roles, motivations, actions, conditions, triggers, sequences, and physical objects. The aim is of course to capture complexity using prescriptions which form recognisable and reusable wholes with an absolute minimum of constraint. It is hard not to get drawn back towards the idea of frames and object-oriented methods, but the more one succumbs to that temptation the less a script will meet the spirit of the original design goal!

It might be easier to scoff a little at such research work as not being very relevant to the workaday world of databases for business information systems and quietly forget about it. But get a business person to talk about work, and the knowledge forthcoming is much more likely to be displayed as a series of scripts than as a bundle of database schema however sophisticated. One would guess that future business information systems earning anything approaching the 'intelligent' label would be sure to contain script-like components of one kind or another. Nevertheless, there is at present nothing approaching a standard script language, and we must await further research and development.

10 Further Knowledge Representation Demands upon Logic

10.1 Introduction

10.1.1 What do we really know?

Our everyday understanding about what we think of as our own knowledge about things is both vague and complex. It is vague because we do not seem to have to be too meticulous about the way we choose to get to know things. Sure enough, if we are faced with specific knowledge acquisition tasks such as learning a foreign language or studying for a specialist examination we do often set up some kind of system for learning, or adopt a recommended system from a teacher or a text, but even then we tend to use our own idiosyncratic version of the system. And no two people will follow an exactly similar regimen. Thus, by the time we have acquired our new knowledge we shall also have adopted with it our own ways of using that knowledge.

For less formal types of knowledge such as knowing how to do the weekly shopping, knowing how to get to work, knowing how to take part in meetings and so on, our 'systems' are often hardly worthy of the name at all since we seem to make use of the knowledge unconsciously for the most part. So if asked without warning to describe in systematic detail how we acquire our knowledge in such cases and how we actually make use of it step by step, almost all of us will fumble and fudge. We need notice of such questions so that if we are to make a serious attempt at answering them we can reflectively set out some kind of system by a combination of guesswork and invention. If anything results from this it is very unlikely to resemble Boolean logic!

Even with the type of knowledge that we might classify as being pure memory work such as we might make use of in playing quiz games, in conversations about TV shows we have seen, in reminiscing, in recalling telephone numbers and names, or in using software packages with many functions and few menus, even with these situations we would be hard pressed to explain how precisely we achieved the necessary recall. We are even less able to explain why we often cannot recall things that we know we know!

And yet all this vagueness is a kind of cover for enormous complexities of mental activity that must in fact take place in order for it to achieve what it does achieve. Our early uses of computers in the business area have understandably not attempted to unravel this mystery for the purposes of getting the machines to help us handle data and information. Instead they have been based on what appears to be the ways in which we used paper-based systems in the past. Files and databases are at heart electronic versions of paper filing systems.

However, the uses of these files and databases have increasingly reminded us of the mental processes that we had to use when looking up paper-based systems and this brings us back to the question of knowledge-using behaviour. Languages for forming enquiries to the data storages systems emerged, which meant that we couldn't then avoid having to think about how we think.

From another direction came the development of systems which deliberately set out to model skilled thinking and decision-making processes. Although the idea of creating intelligence artificially by means of computers and other electronic machines is as old as the business methods mentioned above, it has taken longer for the work to emerge as being seen to be feasible and viable. But it is now happening and expert systems development is very much part of the business scene, though spasmodically and patchily rather than across the whole of business as yet.

For the most part the various attempts to model thinking processes has centred on specially created computer languages which if they use logic at all use it in its simpler forms, and we have already asserted that simple logic is not what we mostly use in our real-life knowledge processes. Researchers are very much aware of this and there has been a growing emphasis on trying to find ways to close the gap. The research is extremely diverse and much of it is difficult to follow in detail, but its aims and directions are beginning to influence software design and we therefore need to find ways of looking at the field which overcome these difficulties if we are to be properly prepared for the developments which are imminent.

This chapter sets out to do just that. Our attempt will no doubt horribly oversimplify the work and will be in danger of offending the specialist. We shall accept that as the price of modestly enlightening the non-specialist.

10.1.2 Three battlefronts

We choose to explain the problems which must be engaged by viewing them from three battlefronts. These are

- The problems that *time* gives logic.
- The problems that *uncertainty* gives logic.
- The problems that *truth* gives logic.

These reflect aspects of real-life thinking that are so common that we may hardly even notice them as being related to logic at all. We think in real time, and time changes all things, so we should ask how this affects our reasoning about things. We are certain about one thing in life and that is that nothing is certain, so how does this affect our reliance upon factually precise data? We know that so-called truth can be false, due to error, temporary assumption or even human capriciousness, so how on earth can we truthfully reason about that?

How indeed? Research does not offer simple answers but the work under way is interesting.

10.2　The problems with time

10.2.1　Introduction

It is not an exaggeration to say that the question of how to handle the time dimension in information systems has largely either been ignored or taken for granted. An example of its being ignored is the common absence of time representation within the data stored in databases, and an example of taking time for granted is whenever the automatic clock or calendar is accessed in a computer system and the data fed into an information system direct. This attitude might arise because we tend not to see that there is any special problem with regard to time. It is a dimension that we all inevitably share and which appears to be relentlessly sequential and therefore well-ordered. But this superficial attitude does not take into account the actual richness of the human mental conception of time—a richness that is reflected, for instance, in the range of verb tenses which we find in natural languages.

To illustrate how this richness lies close to even the simplest of commonplace business events let us imagine the following scenario.

A customer has booked a holiday with a certain packaged-holiday company using a high street travel agency shop. The holiday company keeps the record of this booking on a centralised database, while the travel agency manually files a copy of the appropriate booking form which was filled out from data obtained both from the computer screen and from the customer. Now consider the following time-related questions which could be put some time after this event:

During what dates will the customer be on holiday?
When did the customer first come into the agency shop?
When did the customer confirm the desired holiday?
At what time was the booking recorded at the central database?
At what time was the booking form filed at the agency?
By when must the final payment be made a) by the customer to the agency
　　and b) by the agency to the company?
What are the precedences of these two deadlines?
What rules govern the changing of dates by the customer?

By when should the customer have received flight tickets?

What are the scheduled flight times?

How long before the flight time must the customer check in?

What are the scheduled flight arrival times?

By when must the customer check out of the hotel?

Is the customer on the holiday now?

Will the customer be on holiday tomorrow?

Has the customer cancelled the holiday? If so, when?

What rules govern the customer changing flights?

If the holiday is complete, what were the actual flight times?

If this customer has booked with the company before, when was the customer last on a holiday run by this company?

Do the dates of the previous booking overlap with the dates of the current booking?

If this particular packaged holiday is going to be available next year, what will the period of its availability be?

Although this list of questions is uneven in respect of what we might term commercial value, all questions are ones that might be quite reasonably posed in some context or another. Like any list of potential database queries they raise issues about the scope of the data to be maintained by the particular database system. In that respect they are little different to any other material being used as criteria for the design or evaluation of a particular database.

As a set however they are special in that each question refers, in part at least, to the time dimension. Is this a significant feature or not? Contemporary database practitioners may answer with a resounding No! on the basis that one should only store data that has a reasonable chance of being needed. In any case, dates, times, and periods past, present and future are quite different categories of data, needing different critical attention. A query about a past clock-time event, for example, such as the one above concerning the time of recording the booking, is intrinsically different from one about future dates such as the one about the dates of the booked holiday.

A more theoretically inclined computer scientist may however answer the question whether common time dimension is significant with a cautious "Could well be!" Since current time and date availability is an automatic feature of every computer, one should at the very least ensure that there is compatibility between the data formats of these and the database times and dates.

Knowledge engineers are likely to be even more positive—and therefore radical—in their response since they will want to try to ensure that *all* common patterns of knowledge in the real world—we do agree that time is a common feature of all the above questions, do we not?—should if possible be reflected in some design aspect of the knowledge base. They will especially notice the rule-based questions which begin with If and seem to suggest a need to work with the rules that are able to travel across the time dimension. "If the holiday is completed, what were the actual flight times" is heavy with such 'time travel' implications; today's date, records of completed events both as dates (the holiday) and as times (the flights), and how to handle the future (holiday has started but not yet completed) are all possibly relevant here, with reasoning connections between them being a very likely requirement.

However, there is little in the way of agreement about standard ways to handle time-knowledge representation, and the aim here is make the reader aware of the points at issue and more able to understand the debate about it as the arguments unfold over the coming years.

10.2.2 Modelling time

There are two related aspects to time modelling which are part of everyday experience—what *the time* is (or was, or will be) and how long a *time period* was (or will be).

The instant of time is recordable against a number of types of scale as follows:

Date systems built upon some major historical base line
Week/day systems built upon return-to-unity cycles
Week/day systems built upon day-group cycles
Clock systems built upon return-to-zero cycles.

Date representation presents problems in so far as calendars may vary from one culture to another or, within a given calendar culture, the system of writing dates may vary. Most computer users will have experienced the latter problem at some time or another, most likely because of the fundamental difference in day-month sequence between the USA and Europe, but also possibly due to differences built into the software concerning, for example, whether or not numeric representation is used for months, how to abbreviate month representations when names are used, whether two or four digits are used for representation of years, and whether preceding zeros are used for day representation to achieve fixed-length format. The more sophisticated software packages do nowadays provide for format translation, the actual date being stored by the system in some self-consistent way. It is interesting to note however how many dire warnings have been issued about future system crashes that lie waiting to happen because of the inadequacy of the date format in many older systems—beware the year 2000!

Week/Day (eg. Week No 23 or Day No 156) counting systems are relatively common in business, especially in connection with financial years. Problems arising here include lack of standard for the start of financial year, variations as to which is Week No 1 of the calendar year, and uncertainty as to whether or not non-working (holiday) units are counted. A meter-reading schedule for visits to domestic consumers of gas may be based on a working-day-of-the-quarter, for example, which illustrates this point and also shows how the return-to-unity can be based on other than a full year.

Week/Day grouping can occur as week-of-the-month and day-of-the-week numbering. Day-of-the-week also has a name system attached to it of course! Week-of-the-month systems are inherently confusing because of the length-variation in months. Although human intercourse involving time uses relative time references (tomorrow, a week ago) and the name-day-of-the-week system (Monday, Tuesday, ...) with great frequency, software systems have been slow to build these into their internal calendars. Once they have been incorporated (as they have for example in the Lotus Agenda software), then it is feasible to contemplate a time intelligence being built in to a system, so that phrases such as 'tomorrow', 'in two weeks time', 'one week earlier than completion date', 'last Monday', and so on, can be understood by the software and registered as appropriate dates.

Clock systems are very familiar to us and may therefore seem obvious and simple in their meaning. But this disguises the natural intelligence of humans in dealing flexibly with 'the time'. One aspect is how we cope easily with approximations of time. Whereas dates have a built-in approximation to the day unit (or sometimes the half-day unit), clock systems vary considerably in this respect. Time may be cited to the nearest hour, minute, second or fraction of a second. Fractions of a second are standard within computer systems and the precision may be to several places of decimal. Humans may know instinctively or by context how many places of precision are relevant, but building this into software implies some extraordinary complexity.

The question of relative times around the world's time zones becomes an ever more relevant one as telecommunications bring all these zones into a common real-time frame. This is perhaps one instance where computer systems can prove to be more intelligent than humans and provide intelligent support to continuously functioning international information systems.

However, a further problem with clock systems of any kind is whether or not they are 'right'. They may be fast or slow with respect to some agreed standard and therefore if the time is recorded for future use there is uncertainty about the reliability of such data in the context of different uses to which it may be put.

These issues are very relevant to the second type of time modelling—the length of time period. Accuracy in specifying length of time may be rendered worthless by an inability to be equally precise about start and finish times. This has been a familiar problem in the field of high-volume data entry, where extremely accurate measurements of keyboard operators' speeds and task-time lengths achieved through reference to the computer's internal clock system are made nonsensical by not being able to precisely identify how the tasks fit into the real lapsed 'office' time of the work environment. These different time granularities may need to be matched up—measurements in whole minutes may be more appropriate to the above example.

10.2.3 Events

Perhaps the most straightforward way to approach the question of time as data is to consider it from the point of view of recording a past event. We shall ignore for the moment the nature of the event. A past event has potentially the following time-related facts associated with it:

> A start date and time
> A finish date and time
> A duration (derivable from above)
> A date and time of being observed
> A date and time of being recorded.

These being facts, we can say that they can be treated in logic terms as having values which are TRUE.

If we introduce the possibility of making time-related statements about the event, those statements will be TRUE or FALSE according to how the facts within them agree with the above TRUE facts. We can also introduce a time-now concept which will bring in potential statements about lapsed time since the event, and also logical queries about whether the observing and recording times (events in their own right) have actually taken place. We should remember too that all statement-events have associated occurrence times which may be derived from the time-now of their having been made and recorded.

These observing and recording event times remind us that we very often wish to regard an event as having coincidental start and finish times, that is to say in effect a duration of zero time length. If that is too crude a simplification it can create problems when trying to determine proper ordering of events on the time scale, especially where approximations have also been adopted at too coarse a granularity. Thus business transactions deemed to have taken place 'on the same day' or 'during the same hour' will not be susceptible to time prioritising if there is no finer-grained time data available.

10.2.4 Futures

Events which have not yet taken place, which usually means planned or predicted events in the context of information systems, have similar time associations to past events but with some important differences as follows:

> An expected start date and time
> An expected finish date and time
> An expected duration (derivable from above)
> A date and time of having been planned or predicted
> A date and time of being recorded.

An important qualitative difference is that all the expected times are liable to amendment—at the worst, very frequent amendment. It is likely therefore that the final two lines above represent classes rather than individual items—each could in practice be a string of planning and recording times—and consequently the 'expected' will also be sequences of modified values. Software-based planning systems do not usually maintain a file of all these values, but simply hold the latest one provided.

Because the expecteds are not definite facts there is associated uncertainty about their values, which may require the storage of accompanying confidence-level measures.

Finally, there is the fact that planned events should at some time become actual events when the data then has the qualities described in the previous section. The two types of data—the historical future and the historical actual—should then be linkable in some way. "What was our time prediction performance through the project like?" is a question whose answer requires that such linkage should exist.

10.2.5 Time and truth values

We have tacitly assumed that, when we say that a statement is TRUE, it is somehow always so. What we are usually really saying is that it was TRUE at the time at which it was made. Some statements do admittedly have an assumed TRUE-forever quality. Mathematical statements, for example, seem to be in this category. Only 'seem', because we must always keep the possibility in mind that someone could at some time prove even that $2 + 2 = 4$ is wrong! Indeed the truth of that equation is contingent upon a careful presentation of context and language. Statements about history may also be taken to have this quality of TRUE-forever, in this case with the proviso that it was indeed TRUE in the first place. The circularity is potentially confusing, but again the human mind can usually cope quite adequately with this.

In the real world, especially that of business and industry, accepted truths are changing all the time, so what price the use of eternal logic in such an environment? Even such simple-sounding statements as "The stock level has not fallen to re-order level" can change from TRUE to FALSE so rapidly that the information system of which it is a part exhibits a lag time during which this change of truth value has not registered. The system will reason that stock is available to fulfill an order which may prove not to be the case.

If, however, the system had appropriate intelligence about such lags it could perhaps offer suitable 'health warnings' about the information it was giving about feasible action; for example, "Order provisionally accepted, but please wait two minutes for confirmation."

Traditionally we have coped with this problem of up-to-dateness of facts in systems design by building time fairly crudely into the system. Batch systems are *very* crude in this respect, the data on files being understood to be TRUE at best only up to the time of the last run of transactions. On-line systems are less crude but can offer some very peculiar pictures of the world by referring to data in different files having very varying degrees of up-to-dateness. Data with some kind of time signature attached would help to overcome this problem, but the processing complexity might then be a high price to pay.

However, knowledge system designers are less likely to baulk at this cost barrier, and will want to know why systems should not be able to cope with the effect of time on logic processes. Indeed they broaden the view to incorporate the problem as one of change irrespective of specific time values. There are three ways in which this can be important:

- That truth values change over real time.
- That what may have no truth value to start with (undecided or not captured) may have a truth value later.
- That we may wish to explore possible futures with various new truth values assigned to variables.

Although these all refer to truth value changes, they are sufficiently different qualitatively to justify different approaches. The first has to be concerned with a kind of truth-monitoring system, the second with the infinity of possible new variables that emerging events may bring, and the third with being able to go forward into and back from hypothetical futures.

10.2.6 Sequences in time

Time has unidirectional flow. If that seems too dense a feature to build into our systems, we can say that at least we are able to identify where one time-signatured event should be placed in relation to another. Some event chains are however more strongly placeable in time sequence due to the nature of events. A prescribed task sequence, a progressive dialogue, a set meal menu—all these have time chaining features related to their own content. Note though that the sequence is not immutable necessarily. Operators may choose to change task sequence, conversations may be repeated in different sequence without necessarily changing the total effect, and you may choose to start your meal with pudding!

Most tightly constrained of all are logical sequences which may have only an implicit connection with time sequence. Tasks sequences where each task builds upon the result of a previous task, a chunk of non-branching computer program, and a piece of logically deductive argument are examples. Such sequences are more abstract than others, but when put into operation do still occupy a literal time sequence.

The point about such chains is that time values may be purely relative and indeed could vary considerably without changing the intrinsic nature of the result. Two different operators may take different times over various tasks and still produce the same product (simultaneously even!). A program chunk run on different machines may take different times over each program step. A logical argument is almost timeless in terms of the time needed—think how the same move sequence in chess might last seconds in lightning chess and weeks or months in correspondence chess.

10.2.7   Temporal logics

There have been, and continue to be, various proposed ways of dealing with time representation that will enable a formal logic to work with them. We hope that what has been discussed so far will have raised awareness of what might be the problems facing such attempts. We shall here summarise the main issues.

First, how is the dimension of time to be modelled for the purpose of relating truth statements to it? We could propose an axis which as in any coordinate graph system is used to position points of time against a straight line. This models time as a variable t which could be assigned to any logical variable according to whether we are defining the variable to be TRUE or FALSE. A possible shorthand might be

$$T(t,A) = 1 \quad \text{and} \quad T(t,B) = 0$$

which would represent respectively the assertions that at time t A is TRUE and B is FALSE. We might also want to express the same for time intervals or periods. The variable p might be used to express a time period between two times t_1 and t_2, say. Then

$$T(p,A) = 1 \quad \text{and} \quad T(p,B) = 0$$

would represent respectively the assertions that during period t_1 to t_2 A is TRUE and B is FALSE.

To show how this enables us to build logic around the system we could write:

$$T(t,A) \rightarrow \ {}^{\sim}T(t, {}^{\sim}A)$$
$$T(p,A) \rightarrow \ {}^{\sim}T(p, {}^{\sim}A)$$

which say that if A is TRUE at time t it is FALSE to say that A is FALSE at time t (and similarly for the period p).

$$T(t,A) \ \& \ T(t,B) \rightarrow T(t, \ A \ \& \ B)$$
$$T(p,A) \ \& \ T(p,B) \rightarrow T(p, \ A \ \& \ B)$$

These state that if A is TRUE at time t and B is TRUE at time t, then both A and B are TRUE at time t (and similarly for the period p).

But what about $T(p_1,A) \& T(p_2,B)$? We need some means of symbolising the overlap of the two times p_1 and p_2. Let us define $O(p_1,p_2)$ as meaning values of time t common to both periods. (Notice the similarity to set intersection.)

We now have

$$T(p_1,A) \& T(p_2,B) \rightarrow T(O(p_1,p_2), A \& B)$$

We might well need to have time quantifiers to refer to the special cases of all times past, all times future, at least one time past and at least one time future. Let us adopt $\forall P(A)$, $\forall F(A)$, $\exists P(A)$ and $\forall F(A)$ to denote these respectively. Thus we have

$$T(\forall(A) \& \forall(B)) \rightarrow T(\forall(A \& B))$$

which asserts that if A has been TRUE during all times past and B has been TRUE during all times past then both A and B have been TRUE during all times past. Also,

$$\forall P(A) \rightarrow \exists P(A) \quad \text{and} \quad \forall F(A) \rightarrow \exists F(A)$$

which asserts that if A has been TRUE during all times past then A has been TRUE at least one time in the past and similarly for the future example.

To make for simpler notation we could take the absence of the T function to imply 'all times', ie. past and future. This could apply to mathematical axioms for example. Thus,

$$\forall P(A) \& \forall F(A) = A$$

Also,

$$\tilde{} \forall P(A) \& \tilde{} \forall F(A) = \tilde{} A$$

Note though that this will mean that $\tilde{} A$ must be read as not 'the negation of A' but 'it is FALSE to assert that A is always TRUE', which is not an unreasonable modelling device.

10.2.8 So many other problems

Not all logicians would be willing to accept this kind of modelling as valuable, hence the several different modelling systems that have in fact been proposed at various times. Besides this, there are hosts of other problems to be dealt with, including the following:

- How to represent 'time now': many artificial intelligence applications are concerned with dealing with knowledge as is at a particular moment in time, robotics especially. One might define 'now' as the time squeezed between time past and time future, but how does this affect our modelling of those two features?

- How to model times 'previous' and 'later' relations: this is essentially an ordering (sequence) problem and there are mathematical precedences to work from, but they add a new layer of complexity.
- How to model time 'until' and 'after' relations: many events divide time significantly, especially in connection with the logical chaining mentioned earlier. There is also the phenomenon known as 'persistence' which applies to events that leave the world in a changed state until further changes occur.
- How to cope with alternative futures, which are common in planning for example. Alternative pasts occur in some disciplines too, such as history, archaeology and fault tracing.
- How to model the many verb tenses in natural languages.

All these have been given due attention by researchers (logicians, computer scientists or knowledge engineers) at some time or another, but a comprehensive and standard treatment is still to be found. Meanwhile it may be enough for the general information system designer to use expeditious ways of modelling time according to the application under development and not worry about the general question too much. An awareness of the nature of the general problem is probably useful however, and that is what we have tried to develop here.

10.3 The problems with uncertainty

10.3.1 Introduction

Uncertainty is certainly a problem, isn't it? Well, isn't it? The reader is perhaps uncertain about what is the probable intention behind asking this. But just how uncertain is the reader? Is there even a reader there to experience the uncertainty?

A somewhat cryptic paragraph, but the probability is that with a little perseverance some sense can be made of it and this is indicative of just how robust is our ability to cope with the concept of uncertainty in many forms. Our day-by-day thinking at work or at play copes continually with uncertainty of all kind, and in the business environment there is even a case for saying that the ability to cope with it is one of the more important personal assets.

  We want here to consider how the question of uncertainty is relevant in particular to the uses of logic as a support to knowledge representation. At the base of all Boolean logic we have the law of the excluded middle—a logic variable must be either TRUE or FALSE—which suggests that we must be certain about our data before we can safely submit it to the processes of logical reasoning. On the other hand, predicate forms imply that we could express qualitative truths (and uncertainty could be called a quality) without losing control of the logic processes. Why not express 'unsure of x' as the assertion unsure(x) having value TRUE?

As with the time problem of the previous section, approaches to these questions are diverse and often abstruse. Our aim again is to present the issue in such a way as to help with understanding the nature of it rather than to work through actual proposed systems for modelling uncertainty. We therefore create the following categories of problem under which to achieve this:

Complete uncertainty
Uncertain or 'fuzzy' sets
Uncertain inference

These are not entirely discrete categories and some proposed systems cover more than one of them. But the categories are amenable to progressive treatment in a way which we hope plain common-sense can recognise in the following brief expansions of the ideas.

10.3.2   Complete uncertainty

The reader will no doubt be familiar with the type of conversational response: "To be completely honest with you, I just do not know the answer to that." In setting up files, databases and knowledge bases it is not inconceivable that record spaces will be set up which are empty and awaiting the data to be loaded. If that is the case then this too is like being quite sure that the facts are not (yet) known.

It has been proposed that this state of 'confident' uncertainty could be set up as a third truth value in logic. Let us symbolise it as DK (for don't know). It would seem reasonable to propose that DK should be defined as being strong within conjunction. By this we mean that

$$TRUE \& DK = DK$$
$$FALSE \& DK = DK$$
$$DK \& DK = DK$$

This models the reasonable idea that a conjunction containing uncertainty should itself be uncertain. Doubt does contaminate in this way. In data retrieval terms we are saying that the total result of our search has included an empty record and therefore our result is still uncertain.

The situation with disjunction is more difficult. There does not seem to be any reason why the rule 'at least one TRUE gives a TRUE' should not apply:

$$TRUE \lor DK = TRUE$$

Our search was, say, for any valid data and we did find some.

Against FALSE though, the uncertainty must surely again contaminate the result:

$$\text{FALSE} \lor \text{DK} = \text{DK}$$

Our data search found either non-valid content or an empty record, so our search outcome remains uncertain. In which case it must also be said of course that

$$\text{DK} \lor \text{DK} = \text{DK}$$

With negation it might at first seem right to say that the opposite of 'don't know' is 'do know'! This would not be in the right spirit of our having *confidence in not knowing*. The value, we are saying, would be either TRUE or FALSE if we did know it, so its negation would be either FALSE or TRUE, but we still do not know which. Thus,

$$\tilde{}\,\text{DK} = \text{DK}$$

Think of a data search trying to confirm that specific data was invalid. The data slot in the specified record turns out to be empty. So we still do not know if it's valid.

Uncertainty then almost totally swamps its environment. But not quite. The case $\text{TRUE} \lor \text{DK} = \text{TRUE}$ stands out from the crowd in this respect and this gives powerful justification for believing that this three-valued logic might have some practical effectiveness. We can test for example what happens to implication:

$$A \to B$$
$$= \tilde{}\,A \lor B$$

Rather than set out the whole truth table let us use observation to get at this one. The expression is TRUE only if $\tilde{}\,A$ is TRUE (ie. A is FALSE) or if B is TRUE or if both $\tilde{}\,A$ and B are TRUE; the expression is FALSE only if A is TRUE while B is FALSE; otherwise the expression is DK. Is this reasonable?

Take an example we used in an earlier chapter: If it is raining, the roof is wet. Our new system works like this:

- Either it's not raining or the roof is wet, so the stated implication is valid (the expression is TRUE).
- It is raining and the roof isn't wet, so the implication has failed (the expression is FALSE)
- I am uncertain whether it is raining or whether the roof is wet or not, so I can't tell whether the implication is valid or not (the expression is DK).

The system seems to work for this example at least. It does not of course exempt us from committing ourselves to at least one of the three values for each variable including the don't know! But the problem with truth itself we leave to the final section.

We have assumed that set membership is a Boolean feature in that a specified element is either a member of a specified set or it isn't. "This record is either on this file or it isn't." In the previous section we saw how a don't-know category of truth could allow for saying that one doesn't know whether the record is on the file or not, so we have in a sense covered that eventuality. However, DK is definite—one is saying that one categorically does not have the knowledge—while some kinds of not knowing are in a sense measurable. The probability measure is relevant here.

Take, for example, the 'good customer'. The approach adopted in the previous section would allow a business to classify a customer as being 'good' or 'not good', or they have the option of saying 'don't-know'. But the business itself is more than likely to be able to apply epithets such as 'quite good', 'excellent', 'pretty awful'. 'diabolical' and so on with some confidence. One could of course create a set for each of these categories and establish which set any particular customer belongs to, but then the sense of the implied graduated *scale* of customer quality would be difficult to define in terms of the separate sets even if ordering of the sets within a parent set was introduced.

Proponents of fuzzy sets therefore have suggested that set membership could have a probability value attached which *would* capture the scale. Such scales are very common after all in applied psychology and sociology—all of us must have at some time or another answered questionnaires where we have had to select one of something like 'strongly agree', 'agree', 'neutral', 'disagree', 'strongly disagree' in order to express a qualitative and subjective view as being at a point along a notional scale. As one creates more and more categories from which to choose so one gets closer to a coarse-grained probability measure until a percentage expression has pretty well got one there completely: "I agree 85% with that" (probability 0.85).

The question arises then as to what happens to such measures when sets are combined in logic expressions. Let us refine our 'good customer' a little to show how this might be tackled. Let us say that we determine 'good customer' on the basis of three factors:

Factor	Set
good at Paying on time	P
good at giving advanced Notice of orders	N
good at placing Higher value (and therefore more profitable) orders	H

Suppose there are membership ratings as follows for three actual customers C_1, C_2 and C_3 (shown in brackets after each set symbol):

C_1: P(0.6), N(0.8), H(0.2)

C_2:   P(0.5), N(0.5), H(0.5)
C_3:   P(0.9), N(0.6), H(0.4)

Now, there are various ways in which we might choose to define a customer as belonging to the set G, signifying Good customer. For example:

Minimum membership rating to be 0.5 (only C_1 qualifies)

or

Sum of rating to be at least 1.5 (all three qualify)

or

Multiplication of ratings to be at least 0.125 (only C_2 and C_3 qualify).

The choice must be made pragmatically and can be very difficult to decide upon in practice. A sophisticated expert system might allow the user to select whichever of these (and perhaps others) should apply. A less sophisticated (but easier to use) one might offer only one.

Let us suppose that we have only the first available; we can specify that each rating must be at least 0.5. After all, it is not difficult to imagine that it may be specified as a rule of some kind, such as

IF P($\geqslant$0.5) AND N($\geqslant$0.5) AND H($\geqslant$0.5), THEN C ϵ G

(We shall assume that the reader can read this ad hoc notation without a rigorous presentation and justification of the syntax.)

But what if we then also want to be able to combine the sets in a logic-based way? What are the membership ratings of the composite customers C_1 & C_2 and $C_1 \lor C_2$ and so on? Let us imagine that C_1 & C_2 is the customer resulting from a merger! It makes some sort of sense to rate the new customer on the basis of taking the minimum rating of each pair in establishing the G membership. This becomes

{C_1 & C_2}:   P(0.5), N(0.5), H(0.2)

and customer C_1 & C_2 now fails the test.

On the other hand it might be possible to interpret $C_1 \lor C_2$ as the situation where before a merger we are trying to establish the achievable performance of the two candidate customers and therefore wish to take the better of the ratings in each case. This becomes

{$C_1 \lor C_2$}:   P(0.6), N(0.8), H(0.5)

which passes the G test.

What we seem to be dealing with here is arithmetic using logic as a kind of underpinning. This has strong echoes of the examples given in 5.3.2 and 6.4.3. The theoreticians are in fact very unhappy about calling such work logic at all, and so-called 'fuzzy logic' has come in for much criticism as being founded on shaky theory when it has made an appearance in expert system shells and other software (though sometimes very effectively it has to be said).

Notice that we have in any case taken for granted the answer to the important question of how we are actually to assign the uncertainty ratings in the first place. Are they to be based on statistical analysis, and if so of what quality? Are they to be based upon some expert's opinion, and if so who rates the expert? These are more than just issues of knowledge acquisition because the methods by which the uncertainties are combined and compounded affect outcomes in ways which are themselves open to question. However, many expert systems have been built using techniques of the kinds we have hinted at and it has to be faced up to that real life problems are riddled with such uncertainty. It may be that the answer, as with so much else in systems design, lies in the quality of the testing of the built product. Systems analysts need to have their critical faculties well prepared for facing these questions.

10.3.4 Uncertain inference

In the previous section we attached uncertainty measures to set membership in order to help express the almost inevitable uncertainty attaching to knowledge about the world. But our uncertainty does not stop with 'the facts'. We may also be uncertain about the truth of any rules we specify. In the business world this is often the case with what is considered to be the right decision in a particular circumstance.

Consider the rule:

If delivery address is in Region A, then fuel consumption will be 5% higher. (A hilly region, perhaps.)

and compare this with the rule:

If delivery address is in Region A, add 1% delivery fee.

The first rule is not always going to be true (if ever precisely!), while the second, although related to the first, is presumably always true because it is a standard procedure. Suppose the first rule is observed in practice to be true 83 times out of a 100. We could think about attaching an uncertainty rating of 0.83 to its inferential accuracy. This could be useful in reasoning which is aimed at making decisions about the region concerned, especially when combined with many other similarly uncertain rules.

Again many expert systems offer the user the option of attaching such ratings to rules. The question again arises as to how to handle rule combinations. For instance, what should be the rating of the combined rule:

If A then C (0.8) and if B then C (0.9)

when both A and B are true? Should the likelihood of C be the minimum or the maximum or the multiplication of the two ratings? The answer once again has to be determined by the application in question. But if the expert system offers only one, the multiplication say, what should the designer do? The rueful answer is "be very careful"!

10.4 The problems with truth

10.4.1 Introduction

We might say that we have already in the previous sections been dealing with various problems related to truth. But in this closing section we strike at something even more crucial concerning the ways in which we use 'truth' in our actual thinking. We shall avoid dealing with The Truth with capital T which can be left for religious discussion and the more abstruse reaches of philosophy. What we can show proper concern for though is the question as to the reliability of the notion of truth itself, because all the logic systems we are likely to encounter as systems analysts do take this reliability for granted.

10.4.2 Modes of truth—modal logic

We have already seen that in real-world systems truth is more often than not dependent upon time. A logic system devised to cope with this dependency can be taken as an example of dealing with a mode of truth—a temporal mode in this case—and the logic itself is therefore one type of modal logic.

However, the term modal logic is more often reserved for the perhaps more fundamental twin issues of necessity and possibility. When we consider any proposition and its associated truth value—be it TRUE, FALSE or an uncertainty rating—we are in any real-world situation taking on board a whole swathe of contingencies as to that truth value. For example, the proposition:

Stock level for Part 123 is below danger level

can be taken as a factual result of accessing a database record, but its real truth is contingent upon such things as

Time at which the proposition is stated.
Source of the knowledge (here, the database).
Source of that source (the database maintenance system).
Interpreter of the knowledge (the access program).
User of the proposition.

Thus the message which the 'true' proposition is conveying could actually be any of

Out of date
From the 'wrong' database

The result of miscounted stock
Due to a program error
Misread by the user

and therefore false.

These are all very practical points and should be capable of being catered for by avoidance of system errors of one kind or another. But when we start to treat facts as knowledge, this may not be good enough. We might wish to use the fact plus the contingencies in order to process further down the line. In other words, we may want to have access to more than the one **possible world** which is here represented by the system working correctly in context.

For example, pursuing the above, we might want to know about

The times when danger level was reached.
Comparisons between levels in the databases for two different
  warehouses.
Analyses of miscount occurrences.
Monitored program errors.
Detected user misreadings.

We are asking the system to cater for a greater number of possible worlds, and not unreasonably so given the nature of the example.

Logicians are concerned with possible worlds in a more abstract sense because they wish to develop logic systems which are capable of crossing from one possible world to another without arriving at flawed conclusions. We can easily see that it is not valid to mix reasoning about the similar facts from two different databases, but what if one database represents what person A knows to be true and the other what person B knows to be true about the same real world? Or, to take a different kind of possibility, what if database A is the correct and current database and database B is what might be its state one week from now. We might well exclude from the second the possibility that Part 123 has been recorded as Part 321 but not that its stock level has changed to some new (reasonable) figure.

We shall not attempt here to show any of the ways in which such problems are being tackled since there is still a lack of a standard approach. It is an area which is the subject of much research and it is likely that practical applications in knowledge processing software will begin to emerge in the near future.

10.4.3 ...and as to what really is true...

There are many more questions about the nature of truth values than we have not seen fit to cover here. Two are worth mentioning and make a suitably sobering coda to the book as a whole.

One is exemplified by the suitably named **liar paradox**, which was first referred to in 7.4.1 under the heading "What is a proof?" If a guaranteed hundred percent liar says "I am a liar", then is the statement TRUE or FALSE? We showed that this is actually undecidable as far as basic logic systems are concerned. And yet it is a perfectly simple looking statement and we don't somehow feel that it should be undecidable. In the real world an explanation of some kind can be found—the most obvious being that there is no such thing as a total liar! We should let our uneasiness stand though as a reminder that formalising language into logic is fraught with problems of applicability at every level. Here we have a short understandable statement which cannot even be assigned a truth value because, if we do assign a value, there will be a possibly disastrous bug in the system. If you think it is purely a human example, what about a knowledge base which helpfully contains the rule: "Take every rule in this system to be false"? (Because, for example, there has been a disastrous corruption.)

The other example concerns the potentially infinite nesting of any assertion of truth value. Consider the endless list of statements about the Boolean variable A:

    A
    A is TRUE
    It is TRUE that A is TRUE
    It is TRUE that it is TRUE that A is TRUE
    ...

We might use the first form, for example, in connection with an argument such as

$$A \rightarrow B$$
$$A$$
$$\overline{}$$
$$B$$

where the second line is understood to be read as 'A is TRUE'. But if 'A is TRUE' is understood then so must be 'It is TRUE that A is TRUE' and so on. Therefore, we might as well use A alone in the first place. But is that sufficient? What about its appearance in a general rule such as

$$A \ \& \ \tilde{\ }A = 0?$$

Here A represents either 'A is TRUE' or 'A is FALSE'? The rule however is TRUE for both values of A. It has to be to be acceptable as a general law.

And what about the statement:

    It is FALSE to say that A is TRUE?

This must mean that A is FALSE—but only if the whole statement is itself assumed to be TRUE! (It is interesting to look at the two alternative layouts for K-maps as shown in Fig 2.6 in Chapter 2 and question whether the labelling in (ii) isn't in fact making use of this form of expression.)

This is not simply playing with words. In practical computing terms it is to do with ensuring for instance that a program accessing a set of rules (in a decision table perhaps) is at the right nesting-level setting on entry. As we approach an era when knowledge base systems from different sources may be enabled to interact with one another this could be a less straightforward program design task than it may at first appear to be.

Exercises

Introduction

These exercises are provided not as drill for learning all the techniques outlined in this book but rather as 'self-developable' illustrations of how the techniques might be relevant within systems analysis work. None of them are particularly abstruse or involve much symbol manipulation but neither are they trivial, and it is hoped that by doodling through them as one might a puzzle the reader will enhance his or her understanding of both the simple elegance and the sometimes infuriating shortcomings of first-order logic as a knowledge representation scheme.

They are not graded, but a clue to the ground they cover is given with each. The reader in need of 'classroom'-type exercises should look to more formal and specialised textbooks. Copi has literally scores of small-to-medium length drill exercises, while Frost has a like number but also includes more challenging and discursive questions. Between them they provide a very comprehensive test of knowledge and understanding in the areas of logic and knowledge representation, and as such go well beyond the scope of this text.

1 Boolean modelling

"Either we buy a mini or we buy a local area network system; and we'll have to be careful not to overspend as this implies an inability to handle properly our budget or our purchasing and research policies."

(i) Discuss whether the two 'or', two 'and', the 'not' and the 'implies' in this sentence are intended to be as we have defined them in boolean logic.

(ii) Select suitable symbols for the following, assuming they can be treated as boolean variables:

> "buy a mini", "by a LAN", "spend within limit", "able to handle budget", "able to handle purchasing policy", "able to handle research policy".

Now model the sentence as a boolean expression and simplify if possible. Re-phrase the original in the light of any simplification.

2 Boolean modelling

A certain organisation has a distributed multi-processor system with crucial sub-systems at four locations: London, Birmingham, Manchester and Glasgow. The total system is said to be capable of viable operation if any of the following situations exist:

- All four sub-systems are up and working correctly.
- The Birmingham and Manchester sub-systems are up and working correctly while both London and Glasgow are down for some reason.
- Though Glasgow is out of operation the other three sub-systems are up and working correctly.
- Though London is out the other three are up and working correctly.

(i) Select and define four boolean variables to represent the operational states of the individual sub-systems, making clear what FALSE and TRUE means for each variable.

(ii) Express each of the four situations described in terms of values of the variables.

(iii) Set up a truth table (16 rows) or K-map (16 sub-cubes) for the four variables and identify the rows/sub-cubes which apply in the four situations.

(iv) Using algebraic (from the truth table) or observational (from the K-map) methods derive a boolean expression for the condition of non-viable operation.

(v) Convert this expression into an implication such that you are able to compare it with the sentence: "When there is non-viable operation, the Birmingham sub-system being up and working implies that...."

(vi) This problem is simple enough to be capable of being tackled by common-sense. Check all the steps you have worked through above against such a common-sense interpretation.

3 Boolean modelling

"We always said that we would only purchase the software package if the package proved to be adequately documented, and that we would not in that case hire an experienced systems analyst. Here we are, having hired an experienced systems analyst after all, so what do you make of that?"

Draw up a truth table for the three boolean variables 'purchase software package', 'package adequately documented' and 'hire experienced systems analyst'. By examining the truth table line by line decide what are all the possible explanations of the situation described.

4 Logic modelling of natural language

Here is an excerpt from a transcript of an audiotaped fact-finding interview between a systems analyst (S) and a user (U):

S: So, you would like an A-size screen display, printable in three colours. Actually, your current printer only gives A-size results, doesn't it?
U: Yes. We could use the West London Office printer, just so long as we get the three-colour job.
S: I'll note that. What if that printer doesn't handle A-size? Could you accept that?
U: Mm. Yes...
S: And now, what if three-colour printing just isn't practicable?
U: Oh dear, well, could you then guarantee A-size?
S: Yes, will do. Whichever of the printers we use. OK?
U: Right.
S: In a nutshell then, you'll only accept non-A-size if you are guaranteed three colours, though this could mean not using your current printer.
U: Hang on. That's not what I said.

Is the Analyst's 'nutshell' summary:

(a) correct and reasonable
(b) correct but misleading
(c) just plain wrong?

Use whatever methods you wish to present justification for your selection, but imagine that this is for reading only. In other words it must stand on its own without benefit of your spoken explanation.

5 Analysis of argument

You have been sitting in a meeting trying to follow the reasoning of one of the most fluent of the people present and have decided that the structure of what he is saying is as follows (A through D are propositions that could be true or false):

I   Either II and III are TRUE at the same time or IV is TRUE (or both situations are TRUE).
II   If A is FALSE, this implies B is TRUE.
III Furthermore, if A is FALSE, this also implies C is TRUE.
IV   However, when D is FALSE this makes A being FALSE imply TRUE values for both B and C.

You cause a stunned silence by interjecting with: "Good. So we can neglect D altogether then." Check whether your interpretation is justified by symbolising the above in boolean algebra and simplifying the expression.

6 Analysis of argument

An operations manager is discussing security matters with an auditor. He explains:

"The last working day of the month automatically triggers the system to run the Audit Test program. An attempted entry into the system by an unauthorised user will also do this. There are other events too which are capable of operating this trigger. So you can see that, since today, there has been no instance of the trigger being activated, and it is not the last day of the month, I know that there has been no attempted unauthorised entry."

(i) Is this reasoning as sound as it seems? You can test this by using the special Venn diagram mentioned in Chapters 3 and 5. Draw such a diagram being careful to label it meaningfully.

(ii) Confirm your findings by using algebra.

7 Analysis of rules

A shift manager has devised the following rule to ensure that there is at least one of a three-person shift present in the computer room at any one time:

"If person A leaves, person B must remain in the room, and if person B leaves, person C must remain in the room, and if person C leaves, person A must remain in the room."

(i) Does the system work? Common-sense interpretation shows that it does. How does the logic model show it?

(ii) By expressing the requirement as being A OR B OR C (which is the simplest form of the rule after all), can you see what simple change in the wording to the manager's implication-based rule would make it more precise?

8 Analysis of program decision logic

Working through some old programs for which there is no documentation, you encounter the following statement:

IF (((A) and (not B)) or ((not A) and (((C) and ((D) or (E))) or ((f) and ((G) or (H))))))) GOTO PROCESS-100.

(This has been tidied up from an actual case reported in the computer press some years ago!)

(i) Unravel the logic by setting out the conditional expression in more compact symbolisation.

(ii) Confirm that the branching to PROCESS-100 will always occur if A is TRUE while B is FALSE.

(iii) Experiment with trying to display the logic of the statement as a Venn diagram, a decision tree, and a decision table (or appropriate tableau).

9 Modelling a chain of rules

Here are some statements about program modules kept by a certain department for general re-use:

- If the module is classified as being a smaller program it is never put through rigorous testing.
- If a module is kept on the fixed disk on the departmental central computer it must be of the smaller program class.
- If a module is to be stored as a designated library program it is rigorously tested.
- If a module is designated a non-library program it carries a prefix X in its identity code.

(i) Express each of the statements as a simple implication of the form $A \rightarrow B$, taking care to define each boolean variable first.

(ii) Find an implication 'chain' that starts with "If a program is on the fixed disk..." and uses the information contained within the final statement above. [Remember that $A \rightarrow B$ can always be re-expressed as $\tilde{B} \rightarrow \tilde{A}$.]

(iii) Write out the argument chain in clear narrative form in such a way that any reasonably intelligent person could follow it.

10 Recognising different levels of truth

Below is an 'argument' which you should read through carefully.

"Let us accept as reasonable the assertion that:

He who deduces "There is a 'predicates' section in the program listing, so it must be a program written in Prolog" must himself be a Prolog user.

It now follows that we can assert that:

He who deduces: "There is no 'predicates' section in the program listing, so it must be a program written in something other than Prolog" must himself not be a Prolog user."

The argument is full of problems about questions of material truth, logical deductive argument and changes of level of 'speaker', and is also of very doubtful quality as meaningful argument. Try to identify all these types of problem and indicate how to separate out one type from another.

11 Information as sets

A software director of a large UK company describes the use of various languages in the company's office locations:

"We use different high-level languages across our three sites. Mind you, only C is used at all three sites. Chorley, being our research site, is the only one to use Prolog and Lisp. In common with Cheltenham—our Ministry of Defence contract site—Chorley uses ADA, and in common with our commercial site at Chelmsford it uses PL1. Fortran and Cobol are used at both Chelmsford and Cheltenham. But Cheltenham is the only site now using Pascal. Chelmsford on the other hand is the only site left which uses Autocode."

(i) Express the above information both as a Venn diagram and as a collection of three sets, one for each site.

(ii) Show how the Principle of Inclusion/Exclusion applies for the three sets.

12 Enumeration of sets

"Daily enquiries from our Manchester, Liverpool and Birmingham Offices total approximately 500. Approximately 350 are raised in Manchester and 300 in each of Liverpool and Birmingham. From this you'll realise that many enquiries are raised jointly by two or more Offices in collaboration. If you take all possible pairings of the three Offices, there are about 50 such enquiries for each, and there are about 100 which are raised by all three in collaboration."

Investigate these alleged figures to see if they are self-consistent. If they are not, are the approximations mentioned likely to account for the discrepancy? If not, what other explanations can there be?

13 Analysing decision rules

A loan manager at a bank explains the rules to which he has to work when determining the appropriate loan scheme for a prospective borrower.

"We operate three loan schemes, which we know as SL, ML and LL. These simply stand for small, medium and large loans respectively. We apply five levels of interest rate, termed r1 to r5 which are set according to the status of the applicant and the scheme applied for. (The ranges of the monetary amounts within each scheme and the actual interest rates for r1 through to r5 are updated as economic circumstances change. But the rules stay the same, which is what you're interested in here, I understand.)

Now, rate r1 is for customers of the bank who are in full-time employment, are below retirement age and who are applying for the SL scheme. Rate r2 is for this same scheme where the applicant is not a customer of the bank, but meets the same conditions. This rate is also used for our customers who are retired but have regular part-time employment. Non-customers who meet these conditions will be allocated the rate r3.

Rate r3 is also applied where the applicants are customers requiring the ML scheme and who meet the employment conditions as explained for rate r1. Rate r4 is also for the ML scheme, but this is for our retired customers and non-customers in employment as I described for rate r2. As it happens we also use rate r4 for employed customers applying for the LL scheme. Rate r5 is used only on this scheme and is for retired customers working part-time.

Sounds complicated doesn't it?"

It certainly needs representing differently if we are to be confident of checking our understanding of what the rules actually are. (Advising the bank to rationalise its loan scheme is not part of the exercise!)

(i) First of all try sketching a decision tree with type of scheme as the first branching level and customer/non-customer as secondary branches on each of these.

(ii) Exercise (i) will have made you aware of possible ambiguities in the descriptions of the rule conditions. List these and then decide what assumptions you are going to make about them.

(iii) Now set up an extended-entry decision table to cover all the rules. From this you should be able to determine precisely what the five rates r1 to r5 are used for and also be able to list the combinations of conditions for which the bank does not have a scheme at all.

14 Decision table analysis

The following decision table:

ORDER IS SPECIAL	Y	Y	Y	N	Y	N	N	N
CUSTOMER NEW	Y	Y	N	Y	N	N	Y	N
STOCK AVAILABLE	Y	N	Y	Y	N	Y	N	N
SPECIAL ORDER			X		X			
NON-SPECIAL ORDER						X		X
STANDARD ORDER	X			X				
SPECIAL REPLY		X					X	

has been checked and approved as being correct. Check whether the following structured English is a correct translation of the logic:

```
IF      CUSTOMER NEW
THEN    IF      STOCK AVAILABLE
        THEN    STANDARD ORDER
        ELSE    SPECIAL REPLY
ELSE    IF      ORDER IS SPECIAL
        THEN    SPECIAL ORDER
        ELSE    NON-SPECIAL ORDER
```

The challenge here is to document your checking process in such a way that a reader of your documentation could easily follow your reasoning.

15 Analysis using decision tables

During an investigation into the current usage of PCs in an organisation a systems analyst was handed a copy of the following tabular display:

TYPE OF USE	Word Proc.	Spreadsheet		Data Management		
MANU-FACTURER	All four	All except Apple		IBM Oli-vetti	Am-strad	Apple
ENVIR-ONMENT	All except factory	Per-sonal office	Home use	Personal and communal office	Factory	Home use
SITE A ONLY				yes	yes	
SITE A AND B	yes	yes	yes			yes

(i) Set out this data in a more conventional extended-entry decision table format, using an Else rule column as appropriate.

(ii) Check that there are no overlapping rules.

(iii) Calculate the number of rules in your table and hence deduce how many rules (ie. combinations of the conditions) are subsumed within the Else rule.

16 Logical structuring of rules

The following is an incomplete, relatively incoherent and possibly even incorrect description of a sequence of events to be planned for in designing procedures for a cash-issuing machine. Use the material to test out various techniques—tree, table, tableau, structured English—to identify the shortcomings of the text. Re-present the material in one of the forms in a way which you believe overcomes all the shortcomings. (You may of course have to make assumptions about what was really the intended meaning!)

"The Customer's RESPONSE 1 is to give their ID Number; the card is retained and a warning message displayed if the response does not correspond with the card. Otherwise the Customer RESPONSE 2 is to select one of SERVICE, CASH, ENQUIRY, FASTCASH. SERVICE proceeds to Customer RESPONSE 3 which is to select either of STATEMENT or CHEQUEBOOK. The STATEMENT selection then requires RESPONSE 4, the selection of CURRENT or DEPOSIT, and the appropriate statement will then be printed. The ENQUIRY response leads only to the balance being printed. The CASH response leads to RESPONSE 5 which is specifying the amount required expressed in multiples of 10. The balance in the current account is checked and if OK the requested amount deducted and issued at the machine. If not OK a warning message is displayed and the card returned. The FASTCASH procedure is as for CASH but for a fixed amount of 20 and again the card returned."

17 Use of the quantification symbols

(i) Try to express the following assertions separately in suitable symbolic forms, using ∀, ∃, ˜ etc. as appropriate:

    a All systems analysts have seen a computer.
    b Not all systems analysts have written a program.
    c Some systems analysts have written a program.
    d Some systems analysts have both seen a computer and written a program.
    e A person who has seen a computer is not necessarily a systems analyst.

(ii) Show how, for the assertions in part (i), if *a* to *c* are all accepted as true at the same time:

    a *d* can be derived from these premises (ie. *a*, *b*, *c* as in (i)).
    b *e* can be derived from these premises plus an appropriate assertion that a system analyst is a person!

18 Sets, relations and rules

(i) Try to express the following assertions about the staff in a certain computer services department in suitable symbolic forms, using &, ∀, →, ˜, ∨ and ∃ as appropriate:

    a All staff are categorised under one or two (but no more) of the following job types: manager, systems analyst, programmer, administrator, operator and trainee.
    b Trainees must always be additionally categorised under one of the other job types.
    c There are at present no trainee managers or systems analysts, but there are trainees in the other categories.
    d It is possible that a person may be categorised under any one of the following pairings:

        (i)   systems analyst and programmer,
        (ii) manager and systems analyst,
        (iii) manager and administrator,
        (iv) administrator and operator,

        but no other pairings are permitted (other than those allowed for in *b*).

312

(ii) Imagine that the following relations exist for the above staff:

EMPLOYEE(id_no, name, start_date)
MANAGER(id_no, title, start_date)
SYSTEMS_ANALYST(id_no, title, start_date)
PROGRAMMER(id_no, title, start_date)
ADMINISTRATOR(id_no, title, start_date)
OPERATOR(id_no, title, start_date)
TRAINEE(id_no, title, start_date)

where the identity number (id_no) is company assigned and the start date refers to start date in the job type indicated. The start date with the department is the one held within the EMPLOYEE relation.

What relational operations would need to be performed to list the following information:

a The name, trainee-title and job-title of all trainees.
b The names of managers who joined the department as managers.
c The identity numbers of all systems analyst/programmers.
d The name and job-title of all employees who have only one job title.
e Full details (from all appropriate relations) of employees who joined the department on a specified date?

(iii) The relations in (ii) simply store the data, and this should reflect the effect of the rules implied by part (i). But what if data is added which flout those rules? Think about what this means in terms of the distinction here between the data and the ways of constraining that data to conform to the knowledge of those rules.

19 Knowledge representation

A van-hire company called Cityhire operates its business from a number of English cities—London, Bristol, Ipswich, Birmingham, Sheffield, York and Durham. Each hire centre has a range of models available and customers may rent at one centre and deposit the returned van at another. Van redistribution trips are part of the business therefore, with staff having to travel by other means to fetch vans or to return to home base. A rental agreement applies to one van per customer per rental, though customers may have multiple agreements alive at any particular moment.

A vehicle may be available, on-hire, in redistribution transit or undergoing maintenance. There is a servicing depot at each hire centre and when due for service a vehicle will be serviced at whichever centre it happens to be. Maintenance records are therefore transferred back to the 'owing' hire centre when necessary. The criteria for maintenance being due include vehicle make and model, total mileage, mileage since last maintenance, number of hirings, lapsed time since previous maintenance and, of course, driver or inspector notification of faults. All vans are inspected and cleaned immediately after return by the hirer.

Rental terms are based upon period of hire, model and type of hire. The types of hire are related to whether more than one hire centre is involved in the transaction. Cityhire like to keep track of regular customers by keeping customer files at each hire centre.

Vehicles are all bought new but a van is always sold as soon as it passes its second anniversary with the company.

Data modelling systems analysts will be familiar with this kind of 'information scenario' and the requirement to transform the description into, say, an entity-relationship model. This can certainly be attempted here (see part (vii)), but the intention is that the company's business should first be modelled in a number of ways in order to get the feel of the knowledge implicit in this description.

Here are some tasks to get started on:

(i) Seek out the obvious synonyms and then also the possible synonyms and keep the lists to hand.

(ii) Try free-wheeling through the information in a relatively unstructured way building an informal semantic net as you proceed. This should soon convince you of the richness of all possible connections between words and their associated concepts.

(iii) By contrast, now treat the information as if it can be expected to be expressible as a strict hierarchy. You might find the corporate body of the company at or near the top, for instance, and perhaps nuts and bolts at the leafy ends!

(iv) Go in search of rules of the IF ... THEN kind. Here is a further rule: if necessary then try guessing at rules that could apply in this company, using your previous knowledge and common-sense to induce them! Express these rules in prolog-like clausal-form sentences.

(v) Try to think which parts of the business might be expressible in decision tree (and table) forms.

(vi) Try to think what kinds of things and ideas might be representable as object and what attributes might be assignable to them.

(vii) NOW you can go hunting for entity types and modelling relationships between them. And now it should be clearer just how particular is this view of the business. (Of course, it will be the easiest way to get towards a schema for database design. But who said that only a database was what was required here?)

20 Future directions

Stock control is one of the longest established areas of application of data processing by computer methods. By taking as a specimen example the stock controlling 'intelligence' required in the running of a corner-shop selling newspapers, stationery, sweets and small toys (we can all imagine what this might entail), consider how the 'three battle fronts' mentioned in Chapter 10 might be relevant in considering a knowledge-based support system of the future for this area. This will involve therefore making notes about the possible demands upon the system of time-related, uncertainty-related and truth-related aspects of the shopkeeper's knowledge. Since we have said only little about actual methods of tackling these problems, this is an exercise in 'what' rather than 'how'.

Here are three starter comments by way of example to help trigger the further work:

- Time-related problem: To be able to input customers' newspaper cancellation instructions—'all next week', 'the Sunday after next'—and have the system's screen say what dates are implied.
- Uncertainty-related problem: To be able to interpret tomorrow's weather forecasts as a guide to cold drink stock ordering.
- Truth-related problem: To be able to distinguish between, and keep track of, various types of error that occur with the use of the system, e.g. data, operator, system, customer and stock counting errors.

Bibliography

Bartee (1977, 2nd edn): *Digital computer fundamentals*: McGraw Hill.

Borland International (1986): *Turbo Prolog—owner's handbook*.

Clocksin, W.F. and Mellish, C.S. (1981): *Programming in Prolog*: Springer Verlag.

Copi, Irving (1979, 5th edn): *Symbolic logic*: Macmillan.

Date, Christopher (1981, 3rd edn): *An introduction to database systems*: Addison-Wesley.

Ernst, Christian (ed) (1988): *Management expert systems*: Addison-Wesley.

Frost, Richard (1986): *Introduction to knowledge base systems*: Collins.

George, F.H. (1977): *Precision, language and logic*: Pergamon.

Gray, Peter (1984): *Logic, algebra and databases*: Ellis Horwood/Wiley.

Harmon, Paul and King, David (1985): *Expert systems—artificial intelligence in business*: Wiley.

Hekmatpour, Sharam (1988): *Introduction to LISP and symbol manipulation*: Prentice-Hall.

Hofstatder, Douglas (1979): *Godel, Escher, Bach—an eternal golden braid*: Harvester Press/Penguin.

Humby, E. (1973): *Programs from decision tables*: Macdonald/Elsevier.

Ince, Darrel (1988): *An introduction to discrete mathematics and formal systems specification*: Oxford University Press.

Kowalski, Robert (1979): *Logic for problem solving*: Elsevier North Holland.

Lindley, D.V. (1985, 2nd edn): *Making decisions*: Wiley.

Martin, James and McLure, Carma (1985): *Action diagrams—clearly structured program design*: Prentice Hall.

Montalbano, Michael (1974): *Decision tables*: Science Research Associates/ IBM.

Nilsson, N.J. (1982): *Principles of artificial intelligence*: Springer-Verlag.

Open University: *Graphs, networks and design*: course units for Third Level Course.

Ringland, Gordon and Duce, David (eds) (1988): *Approaches to knowledge representation—an introduction*: Research Studies Press/Wiley.

Silverman, Barry (ed) (1987): *Expert systems for business*: Addison-Wesley.

Stillings, Neil, et al (1987): *Cognitive science—an introduction*: MIT Press.

Stern, August (1988): *Matrix logic*: North-Holland.

Thayse, Andre (ed) (1988): *From standard logic to logic programming*: Wiley.

Turner, Raymond (1984): *Logics for artificial intelligence*: Ellis Horwood.

Turner, Raymond (1990): *Truth and modality for knowledge representation*: Pitman.

Van der Lans, Rick (1988): *Introduction to SQL*: Addison-Wesley.

Woodcock, Jim and Loomes, Martin (1988): *Software engineering mathematics—formal methods demystified*: Pitman.

GUIDE TO THE BIBLIOGRAPHY

The chief aim of this book has been to instruct. A long list of 'difficult' books and papers is therefore felt to be inappropriate. Instead, the Bibliography comprises a carefully considered list of books which it is hoped the reader will be able to use sensibly and without too much fear of encountering unremitting obscurity. There are a handful of books cited which unfortunately may be difficult to track down through being out of print, but larger academic libraries should still have copies of these.

Any author in a specialist subject owes a huge debt to other researchers and writers whose writing has needed to be highly condensed in order to communicate the rapidly developing ideas amongst fellow workers. Such writings are not usually easily accessible to the reader who is striving to learn and are not therefore to be found here. The bibliography does however include a number of books which themselves carry extensive literature reference sections, and these are indicated when appropriate in the chapter-related sections below.

What follows is a brief chapter-by-chapter guide to suggested further book reading. It attempts to provide the reader with opportunities for more direct access to deeper insight into the topics covered.

● *Chapter 1   The Logical Systems Analyst*
We assume that the reader will respond in his or her own way to the ideas

about what types of logical thinking ability is required of the contemporary systems analyst. There is no attempt to offer a book reference on this subject.

However, section 1.2.4 is headed 'Cognitive science' and a suitable text both as an introduction to the basics and for following up the ideas in a way relevant to computing and artificial intelligence is *Stillings et al (1987)*. The opening chapters up to 3 or 4 are sufficient at this stage. All chapters in this text are excellently referenced.

Barry Silverman's own opening chapter in *Silverman (ed) (1987)* entitled 'Should a manager hire an expert system?' is relevant to section 1.1.4. Again many references are appended to each chapter of this useful collection.

- *Chapter 2   Two-valued logics—a simple start* and
- *Chapter 3   Some basic rules of logic*

An important message emanating from both of these chapters is that there are a host of different notations for logic symbolisation scattered throughout the considerable literature and that there are also many ways of approaching the basic ideas. Therefore one hesitates to propel the uninitiated reader towards this problem too quickly and too broadly.

For example, *Copi (1979)* is a respected basic text in symbolic logic, its first edition dating from the 50s, but the reader is taken into the thick of the issues and of the symbolisation quite early in the text and may become daunted if not prepared for the density of ideas. It is a well-tried book however, and the introductory chapter 1 on 'Logic and language' makes a valuably clear entree to its subject.

The first chapter of *Thayse (ed) (1988)*, entitled simply 'Logic', provides a rapid and comprehensive tour of logic from the propositional and predicate perspectives, but it assumes that the reader can cope with logic symbolisation with minimum explanation. The opening chapter of *George (1977)* on 'A background to logic' is succinct, clear and presented entirely without recourse to symbols.

Many of the texts mentioned in connection with later chapters also have their own introductions to logic, usually each with its own flavour. Taken as a collection they provide an invaluable perspective on the multifarious ways of approaching logic, even within the areas related to systems analysis work.

- *Chapter 4   Logical mapping*

The bringing together of most of the ways of 'drawing a picture' of logic operations as we have done in this chapter is, we believe, something of a novelty. The term 'mapping' itself is not often associated in logicians' minds with visual presentation in this way, being more often connected with functional mapping which we have chosen not to address in this text. However, truth tables and Karnaugh maps are often used to explicate logical processes, especially where authors are applying the logic to actual problems. *Bartee (1977)* is a good example of this, where the text is primarily concerned with teaching the basics of logic circuit design.

'The algebra and geometry of logic', chapter 7 of *Montalbano (1974)*, makes particularly interesting reading in the context of our chapter 4, especially as it is part of a text dedicated to the use of one particular mapping type—the decision table (see also under chapter 6 below).

Tableaux and trees are not easily separated-out topics at present as regards further reading opportunities. Technical discussion of trees can be found in any good computer science text, but not often in the sense of being pictorial representations of the logic.

Similarly networks, graphs and matrices are usually part of basic mathematical texts, with little emphasis on their picturing of logic. However, the course units for the *Open University's Graphs, networks and design* are notable exceptions, but are normally only available as part of the course of study. Educational libraries may provide the units as reference-only sources, however, For a glimpse at the as yet not fully fledged subject of matrix logic *Stern (1988)* might be looked into, though it would be challenging material for the non-mathematical reader.

● *Chapter 5   Sets, lists and relations*
A succinct yet useful follow-up on sets and relations will be found in chapters 5 and 6 of *Ince (1988)* and chapters 5 and 6 (coincidentally) of *Woodcock and Loomes (1988)*. Note however that the presentation of these subjects *follows* the introductions to propositional and predicate calculus, unlike the present text. But the contrasting approaches could be taken as instructional.

Further reading on lists, and more specifically the language LISP, can be found in a number of appropriately titled texts. The *Hekmatpour (1988)* is as direct and clear as any, and opens with the interesting question 'Why LISP?'

The same might be said about SQL with perhaps more emphasis since the use of this language seems set to continue to grow over the next few years. Again *Van der Lans (1988)* offers a clear, basic introduction which also has the virtue of discussing the background to SQL at the outset. Chapter 2 giving 'SQL in a nutshell' is useful for a quick-scanning start.

● *Chapter 6   Modelling decision logic*
As a mostly descriptive introduction to the formalising of decision making *Lindley (1985)* makes interesting reading. We have already mentioned *Montalbano (1974)*. It is a very thorough yet readable text on decision tables. *Humby (1973)* gives a fairly straightforward insight into their use in program design. As the date indicates, this is now becoming history as far as software design and development is concerned, but it can be very

instructive history as far as understanding the basic mechanics of implementing decision table specifications is concerned.

- *Chapter 7   Logical reasoning* and
- *Chapter 8   Predicate logic*

Here, we are at the heart of the subject of logic as being of interest in its potential for application. The reader can go in any number of directions from here and the titles of the texts by *Copi (1979, Frost (1986), George (1977), Gray (1984), Ince (1988), Kowalski (1979), Thayse (1988), Turner (1984)* and *Woodcock and Loomes (1988)* give some indication of the choices. As a collection they certainly illustrate the problem the general reader has with having to accept variations in symbol notations.

It is worth saying something in particular about Frost. This 650-page (and more) book is a very remarkable compendium of knowledge about knowledge-based systems. It is worth having to hand as a reference book apart from anything else. The only pity is that it is such a daunting looking book because of both its size and closely set typeface.

Prolog is given a modest introduction in Chapter 8 and further reading should be related to the type of implementation available to the reader. Similar and contrasting introductions will be found in most of the books cited above, but Kowalski's is of special interest since the Prolog language emerges from the book's approach to development of the logic forms and symbolisation. This is deliberate, and Kowalski's work represented an important milestone in the acceptance and use of Prolog.

The *Borland International (1986)* introduction to Turbo Prolog is as specific as one would expect, but is exceptionally user-friendly and instructive for a 'manual'. (We are also indebted to it for the idea behind the Murder Mystery example in Chapter 8.) *Clocksin and Mellish (1981)* is widely accepted as a sound general introductory text.

- *Chapter 9   Knowledge representation*

This is a very rapidly developing subject and it could be invidious to give specific citations since new books will certainly have emerged by the time a reader is reading these words. Again *Frost (1986)* is useful as a reference but *Ringland and Duce (eds) (1989)* is very appropriate as being both a readable and up-to-date collection of papers.

For the more general artificial intelligence approaches, *Stillings et al (1987)* and *Nilsson (1982)* are relevant, the first being especially addressed to the student seeking introductions to the fundamental ideas.

The development and use of Expert Systems must be a very obvious follow-up area and again there is a growing number of texts appearing to cover it. *Harmon and King (1985)* is a well-established text in educational environments. *Silverman (ed) (1987)* is a wide-ranging collection of good-quality papers. Both books are well supplied with references.

● *Chapter 10   Further knowledge representation demands upon logic*
Most of the text mentioned above for Chapter 9 go on in some way to discuss these issues. *Ringland and Duce (eds) (1989)* in particular is recommended. We can also add *Turner (1990)* as a challenging read which tackles the most likely next generation of theoretical representational problems head on.

Though, not entirely relevant to the specific topics of this chapter, *Hofstadter (1979)* is worth recommending for its fine combination of intellectual and entertainment content. It chews juicily at the meat of some of the most intriguing problems that remain with us in connection with logic and knowledge processing, using, as its title promises, music, philosophy and art amongst other things as hunting grounds.

Index